Transformational Sales

Philip Kotler · Marian Dingena · Waldemar Pfoertsch

Transformational Sales

Making a Difference with Strategic Customers

Philip Kotler
Kellogg School of Management
Northwestern University
Evanston; IL, USA

Waldemar Pfoertsch
Business School Pforzheim
Pforzheim University
Pforzheim, Germany

Marian Dingena
MPCN, Action Learning & Business Coaching/
Rotterdam School of Management
The Hague, The Netherlands

ISBN 978-3-319-20605-9   ISBN 978-3-319-20606-6 (eBook)
DOI 10.1007/978-3-319-20606-6
Springer Cham Heidelberg New York Dordrecht London

Library of Congress Control Number: 2015948263

Springer International Publishing Switzerland
© Springer International Publishing Switzerland 2016
This work is subject to copyright. All rights are reserved, whether the whole or part of the material is concerned, specifically the rights of translation, reprinting, reuse of illustrations, recitation, broadcasting, reproduction on microfilm or in any other way, and storage in data banks. Duplication of this publication or parts thereof is permitted only under the provisions of the German Copyright Law of September 9, 1965, in its current version, and permission for use must always be obtained from Springer. Violations are liable to prosecution under the German Copyright Law.
The use of general descriptive names, registered names, trademarks, etc. in this publication does not imply, even in the absence of a specific statement, that such names are exempt from the relevant protective laws and regulations and therefore free for general use.
The publisher, the authors and the editors are safe to assume that the advice and information in this book are believed to be true and accurate at the date of publication. Neither the publisher nor the authors or the editors give a warranty, express or implied, with respect to the material contained herein or for any errors or omissions that may have been made.

Printed on acid-free paper

Springer International Publishing AG Switzerland is part of Springer Science+Business Media
(www.springer.com)

Life is like riding a bicycle. To keep your balance you must keep moving.
Albert Einstein

To Eva and Peter
for eliciting learning of the most profound kind:
giving insight into the essence of unconditional love.
Marian Dingena

I would like to thank all of you who helped me to keep moving in the last years
in particularly Marian and Susanne.
Waldemar Pfoertsch

To Nancy, with everlasting love.
Philip Kotler

Foreword

At Vodafone Global Enterprise, I experience that we are entering a new, transformative era in human and business relations. In particular with our enterprise customers we are driving change and making a difference within the market. Since management and employees want to develop and keep up with demand, we need to understand the unspoken customer needs and transform ourselves and our offerings.

This new book of Kotler, Dingena and Pfoertsch, presents the proven concepts for this transformational journey. They present a perspective and a road map, which guides you and your company or institution into an era of better human and business relations. In the Global Enterprise Division of Vodafone, we are working in line with these principles and experienced astonishing results over the last years.

As shown in one of the many case study boxes of this book, Vodafone has developed a particular relationship to our strategic customer Amazon, in a joint go-to-market strategy. Since Amazon wants to increase their customers' options with an 'always on experience', Vodafone happily took the challenge to make that happen. Amazon also wants 'to be the earth's most customer-centric' and gave us the opportunity to transform ourselves with them to deliver the most outstanding performance to their customers.

To work with these kind of forefront customer companies can be an exciting undertaking, and it may at the same time be a lengthy learning process. The vision and best practices shared in the following chapters may provide insight to undertake this journey in a smoother way. It starts off in Chap. 2 with separating Strategic Customers from others, and you will find the definitions and various management tools to determine the future value of your customer portfolio. Our experience shows that it is not easy to move from transactional and solution sales to transformational sales. You need to balance risk and opportunities internally and externally. You need to have your financial implications constantly on your screen and you need to know which perspective you have with the single customer and in your overall situation. It is important to realize that the transformation includes your entire workforce from sales to solution design and service, but also all supporting functions such as legal and HR need to be part of the journey. The authors have put together all available insights and many examples to illustrate the starting process for your transformation.

In Chap. 3, the authors take you on a ride to advance your sales to a strategic level, moving it away from the actual selling of existing products and services to engaging in business conversations with customers and other parties in the value network. They provide you with the principles of the Extended Decision Making Unit and give instruments for developing customer insights with the Supplier Adaptive Capabilities grid. They encourage you to develop a Joint Strategic Focus and discover Value Innovation Opportunities.

The book offers many illustrative real life examples such as Apple launching the iCar platform, Shell New Lens Scenarios, the 'Embedded engineers' concept as applied by Festo at Marel and Europcar and Daimler who launched the car2go mobility system. Other examples are from Philips, ABInBev and GE.

At Vodafone we know that guiding business transformation (Chap. 4) needs to acknowledge the intrapreneurial role and mindset in the company. I agree with the statement in the book from Ivo Rook, Director Northern Europe at Vodafone Global Enterprise, that 'real value is created beyond systems and processes'. This is a requirement to successfully drive change with strategic customers. If you can guide 'company-altering' or 'behavior-altering' experiences with your employees and customers, the chances are high that you can mutually increase customer value.

Guiding business transformation is making customers more competitive and successful in their markets. This means for you that you need to act as a lead collaborator with them in whole global value networks. You need to offer business-altering value propositions, which could be very specific, even tailored for one customer. This could mean success for you and the customer, i. e., reciprocal value, creating a win-win situation.

This book describes many examples of business transformation: Dell – streamlining processes by moving Unilever's IT deployment into Dell's factory, Europcar moving your way – flawless experience for business travelers, Festo reducing Total Cost of Ownership at strategic customers, GE Aviation increasing 'residual value' for Boeing Business Jet customers. Other examples are about LSI Logic Corporation, Kodak, TNT and our before mentioned Amazon case study.

I could not agree more with the statement in the book (Chap. 5) that the most powerful way to increase impact may be by inspiring others to release untapped potential. Starting with a new mindset, a shift in focus is required from 'what we need from others' towards 'what they may be capable and willing to contribute'. We learn what we need for creating alignment around vivid and factual business opportunities. You and I know that this is not an easy task, but we can learn from the examples given in the book about world renowned companies like Siemens, Electrolux and TNT. TNT Express example – as one of the world's largest express delivery companies with a global reach – describes how customer insight develops into solid business development and how they create alignment and deliver the promise.

In Chap. 6, the book rounds up the efforts of undertaking the transformative journey. We at Vodafone know that Transformational Sales requires disrupting both the customer's and the supplier's thinking and assumptions about their business, but we also know that the payback is worth doing it. When you can create a learning partnership at all touch points in your company and have a shared improvement agenda with the client, you are learning real-time. This puts you ahead of the curve of the competition. It did it for Vodafone[1] and can do it for you. This book gives you all the ingredients for the transformation.

Jan Geldmacher, CEO Vodafone Global Enterprise

---

[1] Note of the Vodafone Group: Note that Vodafone is a trademark of the Vodafone Group. Other product and company names mentioned herein may be the trademarks of their respective owners.

Preface

In the globalized world more and more companies are challenged with ongoing pressure to generate greater revenues from existing customers. The vital ones expand into new and emerging markets and maintain healthy profit margin. Many leading organizations are at a cross road; they are looking to accelerate the performance of their sales teams and they are empowering the marketing organization. In the meantime customer needs are changing and the employees are recognized as an important, game changing asset of the organization. Not all companies master this array of conflicts and drift from their customers and people.

With the concept of transformational sales, firms and institutions have the opportunity to master this challenge. Transformational sales brings commercial thinking to a new turning point, incorporating human and organizational requirements into a concept which will allow high potential sales and marketing leaders to accelerate growth and achieve break-through success with new organizational setup and process thinking. This holistic approach can help executives to improve sales performance by honest interaction with customers and employees. Transformational sales can help realigning the strategic resources, optimizing sales operations, harnessing sales talents and providing a framework to maneuver through turbulent times.

This is not about reaching the quarterly numbers; it is about driving the business to new heights. With no doubts, it will be a struggle to be heard in the organization bringing dramatically new value to the customer. But it is worth to avoid price wars and to break out of the functional silos and broaden your impact and to increase value to your and to your customer's bottom line.

Transformational sales is changing the thinking of how to do business by sledging away the compartments and boundaries between the functional silos. By transforming your company from 'a product and serice centric company' to an 'adaptive capability centric company' it is necessary to see also the human part of any business transaction.

Henry Ford founder of the Ford Motor Company provided the world with the assembly line technique for mass production. Many companies around the world followed his principles and provide ample products in many categories. Transactional selling process was the basis of their success.

A next layer of marketing processes appeared when companies like Proctor & Gamble introduced systematic concepts of market analysis and marketing communications. First forms of brand management appeared but still the main focus was on transaction based thinking.

With persons like Steve Jobs and the increased use of the Internet, customer engagement moved into the foreground of any marketing thinking. Communication between customers and companies become instrumental to almost any business. One of the goals was to provide seamless and consistent customer experience.

But now we have reached another layer where transformation thinking moved into the foreground. Value is created for the customer by empowering employees and enabling organizations to transform themselves.

Transformational sales is the term and concept that leads companies to the next level. Companies like Vodafone, ABB, IBM and Bombardier have proven that a transformational approach can make a difference, a difference for their customers, their organizations and their employees.

In this publication we provide striking insights with multiple examples, which demonstrate that thinking in transformational sales terms has an impact and is here to stay. Successful examples of companies like GE, HP, P&G, and SAP illustrate various steps in the following chapters. We also provide a transformation agenda where we guide business leaders, academic scholars and students through the journey of transformational sales.

This concept is one of the next steps of marketing science development and builds on Marketing 3.0 with its holistic approach and human touch. It will drive change and guide business transformation for a better world.

Philip Kotler, S. C. Johnson & Son Distinguished Professor of International Marketing, Kellogg School of Management, Northwestern University, Chicago, USA

Marian Dingena, MPCN Action Learning & Business Coaching/Rotterdam School of Management, The Hague/Rotterdam, The Netherlands

Waldemar Pfoertsch, Professor International Business, Pforzheim Business School, Pforzheim University, Germany

Abbreviations

B2B	Business to Business
B2C	Business to Consumer
CCO	Chief Commercial Officer and Chief Customer Officer
CPO	Chief Procurement Officer
CRM	Customer Relationship Management
DESTEP	Analysis of Demographic, Economic, Sociocultural, Technological, Environmental and Political changes and developments
DMEDI	Define, Measure, Explore, Develop and Implement: a conceptual approach to the design of new processes based on the analysis of customer needs
DMU	Decision Making Unit, also referred to as Buying Center
EDI	Electronic Data Interchange
EDLP	Every Day Low Pricing is a pricing strategy adopted for example by Wal-Mart and Procter & Gamble promising customers a low price without the need to wait for a sale price or comparison shop. EDLP saves retail stores the effort and expense needed to mark down prices in the store during sale events
FMCG	Fast Moving Consumer Goods
LAN	Local Area Network: Lans are capable of transmitting data at very fast rates within limited distance
LEAN	In essence doing more with less. Lean manufacturing involves efforts to eliminate or reduce activities which do not add value. Originally developed by Toyota and used in manufacturing, spreading to other functional areas. Popularized by Womack and Jones in their book 'Lean thinking' (1996).
QLTC	Quality, Logistics, Technology and Costs: a model used by ASML to improve collaboration with their supply chain
SAMA	Strategic Account Management Association
TCO	Total Cost of Ownership
TERP	Top Executive Relationship Process: term used within Siemens in relation to the executive engagement process in strategic sales
TRACK	Trends, Relationships, Alignment & Create, Knowledge: entrepreneurial sales approach implemented by TNT

Contents

1 Introduction 1
   References 6

2 Driving Change with Strategic Customers 9
   2.1 Value of Customers: Do They Make us Change? 11
   2.2 Value to the Customer: Are They Willing and Committed to Change with us? 17
   2.3 Transformative Relationships: Driving Change! 32
   References 37

3 Setting the Joint Transformation Agenda 41
   3.1 Customer Insight: Customer Business Relevance 42
   3.2 Company Insight: Supplier Adaptive Capabilities 55
   3.3 Joint Strategic Focus 62
   References 67

4 Guiding Customer Business Transformation 71
   4.1 Making Customers More Successful in Their Markets 72
   4.2 Acting as a Lead Collaborator in Global Value Networks 87
   4.3 Business-Altering Value Propositions 89
   References 97

5 Enabling Internal Transformation 101
   5.1 Building Your Internal Network: The Inside Selling Role .. 102
   5.2 Creating Alignment Around Vivid and Factual Business Opportunities 109
   5.3 Impact: Leading From Any Chair 114
   References 118

6 Undertaking the Transformative Journey 121
   6.1 Creating a Learning Partnership at All Touch Points 122
   6.2 Paving the Path as You Walk on It 124
   6.3 Ending Where It All Begins: Challenging the Own Assumptions 132
   References 133

Appendix 137

Credit Lines for Permission Clearance 147

About the Authors 151

Bring Us in to Speak at your Next Event 155

Index Key Words 157

Index List of Companies 161

List of Figures

Fig. 1.1	Transformational sales: five building blocks	5
Fig. 2.1	Examples of indicators to assess potential value of customers	12
Fig. 2.2	Separating strategic customers from others	16
Fig. 2.3	Match or mismatch?	16
Fig. 2.4	International Purchasing Survey: Purchasing Ratios across industries. (IPS Data 2009, provided by Finn Wynstra, Rotterdam School of Management, reproduced with permission)	18
Fig. 2.5	Dupont Analysis Heineken NV (2014): Impact of purchasing savings on Return on Capital Employed. (Van Weele 2014, p. 13, updated by Van Weele with data 2014 in April 2015, reproduced with permission)	19
Fig. 2.6	Six stages of purchasing maturity and related purchasing focus. (Based on Van Weele 2014, p. 68, reproduced with permission)	20
Fig. 2.7	Customer perspective: Kraljic purchasing portfolio	26
Fig. 2.8	Typical purchasing portfolio for an automotive company	26
Fig. 2.9	Customer perspective upon supply: reverse purchasing	27
Fig. 2.10	Differentiated customer strategies	33
Fig. 2.11	Changing focus: towards transformational sales	35
Fig. 2.12	Changing the customer's and supplier's thinking and way of doing business	36
Fig. 3.1	Insight into upcoming customer business challenges	43
Fig. 3.2	BCG's value creators report: The global population is increasingly connected (Source: Boston Consulting Group 2013; reproduced with permission)	44
Fig. 3.3	Porter's 5 Forces driving competition within the customer's industry	48
Fig. 3.4	Extended Decision Making Unit (DMU) of a supplier of food ingredients	52
Fig. 3.5	Customer Insight: vision on untapped business potential	54
Fig. 3.6	Entering the conversation at a preliminary stage with selected change agents	54
Fig. 3.7	Company Insight: Supplier Adaptive Capabilities	56
Fig. 3.8	Customer-supplier collaboration matrix: value innovation opportunities	63
Fig. 3.9	Joint strategic focus	64
Fig. 4.1	Making customers more successful in their markets: four windows of opportunity	74
Fig. 4.2	Business impact: eight ways to make a difference to the customer business	77
Fig. 4.3	Moving to nonlinear value networks: example ICT	88
Fig. 4.4	Three ways in which value propositions are conveyed	90
Fig. 4.5	Business-altering value propositions	91

Fig. 4.6	Win-win: Thomas-Kilmann Instrument in customer-supplier interaction (supplier perspective)	96
Fig. 5.1	Business transformation requires an integrative perspective on the sales role	104
Fig. 5.2	Benefits of top executive engagement to Siemens and their strategic customers. Source: Senn, 2006, p. 33, reproduced with permission	108
Fig. 5.3	Impact of Top Executive Relationship Process (TERP) at Siemens: The Executive Growth Factor, Source: Senn, 2006, p. 34, reproduced with permission	108
Fig. 5.4	Strategic internal relations may provide access to cross boundary relations	109
Fig. 5.5	Joint Profit & Loss	112
Fig. 5.6	Strategies for driving internal change	116
Fig. 6.1	Relationship experience: learning at all touch points	123
Fig. 6.2	Challenging the own assumptions	132

List of Exhibits

Exhibit 2.1	Characteristics of genuine business relationships	13
Exhibit 2.2	Assessing the customer relationship. (Based on Senn (2012, p. 38), reproduced with permission)	14
Exhibit 2.3	Joint Innovation with strategic automotive customers at Kendrion. (Based on interviews with Dr. Bernd Gundelsweiler (CEO Division Automotive) and Piet Veenema (CEO Kendrion) in September 2013, published with permission)	15
Exhibit 2.4	Sourcing transformation at IBM	19
Exhibit 2.5	Inventing the 21st century purchasing organization (McKinsey survey of >200 CPO's)	23
Exhibit 2.6	Strategic collaboration with suppliers at Bombardier Transportation	23
Exhibit 2.7	Top 10 future hot topics in Purchasing and Supply Management	24
Exhibit 2.8	Reverse marketing at Sony	24
Exhibit 2.9	Deepening the understanding of purchasing strategies: include competitive priorities. (Source Ateş (2014), and interview with Melek Ateş Mach 2014, reproduced with permission)	31
Exhibit 3.1	Connected travel (Columbus 2014)	45
Exhibit 3.2	Industry 4.0: the fourth industrial revolution is already on its way (Roland Berger 2014, p. 7–9; reproduced with permission)	47
Exhibit 3.3	Blurring industry borders: Apple launches iCar platform (CNBC News 2014)	49
Exhibit 3.4	Royal DSM: Customer Insight means 'thinking B-to-C and acting B-to-B' (based on interviews with Mauricio Adade (Chief Marketing Officer DSM), Theo Verweerden (Marketing Program Director Value Creation), Rossana Rodriguez (Senior Marketing Consultant, DSM) in November 2014; DSM 2014a, 2014b; Levi Strauss & Co 2014; published with permission)	52
Exhibit 3.5	New Lens Scenarios at Shell (Extract from Shell 2013; reproduced with permission)	53
Exhibit 3.6	Festo: 'Embedded engineers' at Marel (based on interviews with Folkert Hettinga (Industrial Sales Manager Food & Beverage, Agriculture at Festo), April 2014, and Festo Highlights 2014; published with permission)	57
Exhibit 3.7	Europcar and Daimler: car2go – on-demand mobility (based on interview with Esther van Koot (Commercial Director Europcar Netherlands) in May 2014 and Europcar Activity Report 2011–2012; published with permission)	58

Exhibit 3.8	Philips: applying natural daylight simulation technology in promising areas (based on interviews with Selin Kelleci-Van Balen (Senior Regional Product Marketing Manager at Philips Lighting), Matthew Cobham (Lighting Application Team Manager, Indoor Professional Lighting Solutions Europe), June 2014; Philips 2013a, 2013b, 2014; published with permission)	60
Exhibit 3.9	ABInBev and JF Hillebrand: redefined value in Global Beverage Logistics (based on interviews with Pierre Bonel (Chief commercial Officer) and Sander Ouwehand (Corporate Accountmanager), December 2013–April 2014; published with permission)	62
Exhibit 3.10	Four perspectives on joint innovation (Kim and Mauborgne 1997, 2005)	65
Exhibit 3.11	GE's Quest Program: 3D Printing Quest to improve efficiencies in healthcare industries (GE 2014)	66
Exhibit 4.1	IBM: offering business partners an opportunity to contribute to something big (Moss Kanter 2008, p. 44)	74
Exhibit 4.2	Dell: streamlining processes by moving Unilever's IT deployment into Dell's factory (Biemans 2010, p. 113)	75
Exhibit 4.3	Europcar moving your way: flawless experience for business travelers (based on interview with Esther van Koot (Commercial Director Europcar Netherlands) and Europcar Activity Report 2011–2012; published with permission)	76
Exhibit 4.4	Trust Equation (Maister et al. 2000)	81
Exhibit 4.5	Festo: Reducing Total Cost of Ownership for their Global Customers (based on interview with Folkert Hettinga (Industrial Sales Manager Food & Beverage, Agriculture at Festo), April 2014, published with permission)	83
Exhibit 4.6	GE Aviation: increasing 'residual value' for Boeing Business Jet customers (GE 2013)	84
Exhibit 4.7	LSI Logic Corporation and VLSI Technology: enabling customer's customization (O'Cass and Ngo 2012, p. 133)	85
Exhibit 4.8	Kodak: accelerating time-to-market for consumer goods producing companies (Kodak 2009)	86
Exhibit 4.9	Joint Go-to-Market: Vodafone and Amazon to increase 'always on experience' (Source: Vodafone 2014 – Vodafone Global Enterprise Amazon Case study, published on Vodafone website; reproduced with permission)	86
Exhibit 4.10	Procter and Gamble: joining forces with competitors to improve supply chain efficiency of retailers (Wilson et al. 2001, p. 73)	89
Exhibit 4.11	Value-bridge at TNT: design a close to damage free process (based on interview with Hugo Koppelaars, Director Sales TNT, February 2013; published with permission)	93
Exhibit 4.12	Value proposition of ICT-supplier to strategic healthcare customer	95
Exhibit 5.1	Top 7 Challenges in Strategic Sales practice (Dingena and Teven 2015)	102

Exhibit 5.2	From Customer Insight to solid business development at TNT (based on interviews with Martijn Legemaat, Corporate Account Insight Director at TNT, June 2013–January 2014; published with permission)	103
Exhibit 5.3	Connecting People (Helsing et al. 2003, p. 53)	105
Exhibit 5.4	Orchestrating customer-supplier interaction at Siemens (Siemens Annual Report 2013; Yip and Bink 2007b, p. 13)	106
Exhibit 5.5	Leverage of established relations within hospitality group in moment of truth (Senn et al. 2013, p. 37)	107
Exhibit 5.6	Intrapreneurial role and mindset (Helsing et al. 2003, p. 21)	110
Exhibit 5.7	Electrolux Profit and Loss Statement for strategic retail accounts (Bailey and Hesselschwerdt 2006)	111
Exhibit 5.8	Creating alignment and delivering the promise at TNT (based on interview with Swinda Hagedoorn, Director Global Solutions Management TNT, June 2013; published with permission)	113
Exhibit 5.9	What does it look like when you have impact inside your own company? (Helsing et al. 2003, p. 20)	114
Exhibit 5.10	The silent conductor (Zander and Zander 2000, p. 69)	117
Exhibit 5.11	How much greatness are we willing to grant? (Zander and Zander 2000, p. 73–74)	117
Exhibit 5.12	Listening for passion (Zander and Zander 2000, p. 74)	118
Exhibit 6.1	Vodafone the Power of Simplicity (based on interviews with Ivo Rook, Director Northern Europe at Vodafone Global Enterprise, April to September 2014, and Vodafone (2013, 2014); published with permission)	127

1
Introduction

> *When giants transform themselves from impersonal machines into human communities, they can transform the world*
>
> Moss Kanter (2008, p. 44)

In a globalizing and connected business world, sales has become more strategic than ever before. Successful companies such as ABB, IBM, Bombardier, GE, HP, P&G, SAP, and Vodafone have proven that the value of the customer base is a strong indicator for company value. In this respect several authors have argued that strategic customers should be managed as assets and therefore time and money spent should be regarded as investments rather than expenses (e. g. Ingram et al. 2002; Gupta and Lehman 2005; Tarasi et al. 2011; Senn et al. 2013). We believe that joining forces with strategic customers can be a driving engine to realize change and sustainable market advantage. It is a power and inspiration for challenging existing business assumptions and creating new perspectives on the marketplace, by rethinking the market and business to realize joint value innovation and increase value across traditional company borders.

We are entering a new, transformative era in human and business relations. As argued by Pine and Gilmore (2014) we are moving towards a 'promising frontier in the dynamic experience economy' by creating 'life-changing or company-altering – transformative experiences'. From a marketing angle this transformational era is referred to as 'Marketing 3.0' by Kotler et al. (2010). With Marketing 3.0 Philip Kotler initiated a paradigm shift: whereas the focus of traditional marketing (1.0) was on the bare product, marketing 2.0 has increasingly taken into account the perspective of the customer. Kotler refined this rather one-dimensional model into a holistic approach, and understands the customer as what he/she is, after all: human! People are multi-dimensional, people have different needs and people take every day a large number of different roles in different groups. Marketing 3.0 was born, and the need for companies to adapt to it.

Piercy (2010, p. 350) suggests in his description of the evolution of the strategic sales organization, that a 'radical transformation' in the traditional role of sales is taking place in many companies, 'putting sales back on the boardroom agenda'. Indications of a radical change of the sales role have been visible for some time already, referring for instance to Ingram et al. (2002, p. 559) who declare that 'the sales function is in the midst of a renaissance – a genuine rebirth and revival. Progressive firms are becoming more strategic in their approaches to sales'. In this line of reasoning, in the 'Challenger Sale', Dixon and Adamson (2011) advocate the transformation of sales conversations to move beyond solution selling towards insight selling.

Within this book we provide a perspective upon the possible business transformation that may result from transformative collaboration between suppliers and their strategic customers. Transformational sales is about guiding change and enabling business transformation within this new area, or as Adamson et al. (2013) phrase it, about 'disrupting the customer's thinking and assumptions about its business'. We would like to add that in practice this requires disrupting the supplier's thinking and assumptions about its business as well. In essence transformational sales will transform both the seller and buyer businesses. Transformational Sales is all about guiding organizational change and business transformation, starting with sales itself. Companies using this approach can make a meaningful difference, moving beyond the competition.

Prerequisite to building transformative relations is the selection of a limited number of true strategic customers (see also Cordón and Vollmann 2008). This requires separating strategic customers from other large or 'major' customers (e. g. Rackham and De Vincentis 1999; Yip and Bink 2007a; Piercy and Lane 2009). Besides being important in terms of current value and potential growth, strategic customers make us change and are at the same time willing and committed to change with us. They are the ones with whom we will drive business transformation.

Shaping the joint future and reinventing both the customer and supplier business, requires a joint transformation agenda. This includes envisioning new futures and the willingness and courage to take calculated risk to make them happen, jointly setting the strategic focus based on prioritized value innovation opportunities, and disrupting existing ideas and assumptions about the current way of doing business. This starts with Customer Insight, beyond expressed customer needs. In order to inspire customers with new insights into their business and making them more competitive and successful in their end user markets (Piercy 2010), a deep understanding of disruptive market trends and the impact on customer business challenges is required. This involves an understanding of how the customer world is changing and how that will impact the customer's major business challenges and business headaches and allows identifying the most relevant, prioritized issues to initiate conversations with selected stakeholders within the customer organization.

To be able to connect the dots, strategic sales teams also need to understand their own business in great depth. In addition to Customer Insight, strategic sales teams need Company Insight, meaning a profound insight into their own company's available and accessible resources and competencies beyond current products and services sold. The focus lies on the 'adaptive capabilities' (Day 2011) required to make the difference. This includes accessible resources within the value network. In joint strategic sessions between customer and supplier teams, joint value innovation opportunities can be discovered and joint strategic priorities can be chosen. This results not only in the joint transformation agenda for the coming years but in addition into a 'mental contract' between both firms.

As highlighted by Ateş (2014), customers may set different 'competitive priorities' within their operations and purchasing strategies to increase their competitive advantage in their markets. Based upon the customer's competitive priorities and the supplier impact in the customer organization, we distinguish four perspectives upon increasing customer's competitiveness. Within this framework we elaborate eight ways ('business-altering experiences') to guide the actual business transformation and make customers more successful in their markets. The actual transformation may be the result of the collaborative action between customer and supplier. In many cases it also involves other parties or alliances in the cross boundary value network. Suppliers may become a 'lead collaborator' (Vitale et al. 2011) in an integrated (global) value network. To visualize and vocalize the essence of what the increased competitiveness will look like and building upon the earlier work of Anderson et al. (2006, 2008), we provide an aid to craft personalized 'resonating' business altering value propositions.

Transformational sales is not limited to strategizing with the external customer, it also requires internal sales and network building. The internal selling role (e. g. Speakman and Ryals 2012) is of vital importance to mobilize required resources and competencies and to enable internal transformation. Enabling internal transformation requires a mindset that selling internally is of equal importance *and fun* as selling externally. This integrative perspective on the sales role involves broadening the analysis of the Decision Making Unit beyond company borders, including stakeholders in both the customer and supplier organization.

It also requires an *intra*preneurial role and spirit within the strategic sales team to essentially become an entrepreneur within the boundaries of the organization. To be successful, transformational sales programs need to be regarded as a business rather than as a sales initiative (e. g. Sherman et al. 2003), to create alignment around vivid and factual business opportunities, supported by solid business cases, including joined Profit and Loss statements. We believe the most fundamental aspect of enabling internal transformation is the (explicit or implicit) perspective strategic salespeople may have upon realizing change and impact in the own organization.

Building upon the work of Quinn (1988, 1996, 2004), we distinguish four perspectives on leading change within the organization: convincing, imposing, bridging and inspiring. In our experience the most powerful impact results from the inspiring strategy which is referred to by Zander and Zander (2000) as 'leading from any chair', empowering colleagues to realize their full potential and actively inspiring colleagues to contribute to their best to the new business opportunities. Or, as phrased by Benjamin Zander, a music conductor of the Boston Philharmonic, becoming the 'silent conductor' listening for passion and commitment, thereby inspiring people to release untapped potential.

Transformation of the customer-supplier interaction does not happen automatically. It is a process of leading change. It can be regarded as a 'transformative journey' (Johnson and Fillipini 2009 as cited in Wießmeier et al. 2012). In practice the customer and supplier will go through different stages of collaboration (Cordón and Vollmann 2008), before they enter a stage of strategic alignment. This requires a learning partnership at all touch points in the customer-supplier interaction. The joint transformative journey entails calculated risks and dealing with uncertainty. To a certain extent it is a journey into the unknown which requires to 'pave the path' or to 'build the bridge' as we walk on it (Quinn 2004).

We will end this book where it in fact all begins. We believe that guiding change starts with a conscious reflection upon the own (implicit) assumptions that strategic salespeople may have. Challenging (some of) these assumptions may be the real starting point for driving change. We provide an overview of some of the assumptions that we observe in practice and that may be challenged to drive change.

Five Building Blocks
This book explores how to transform sales in order to transform the customer and supplier business. To inspire both the customer and supplier organization to alter their behavior and change the way of doing business, we distinguish five building blocks (see Fig. 1.1).

- *Building block 1: Driving change with strategic customers.* Transformational sales starts with the careful and conscious selection of those customers with

1 Introduction 5

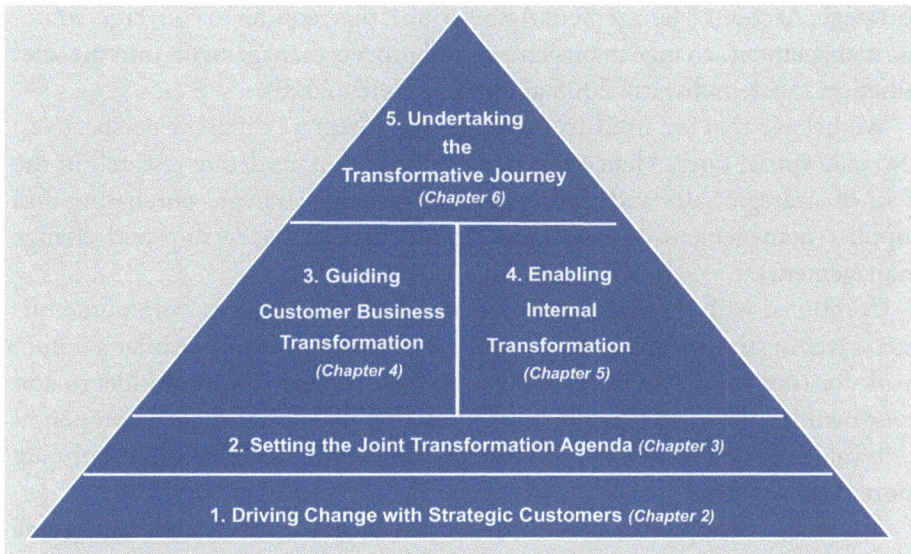

Fig. 1.1 Transformational sales: five building blocks

whom we will be able to drive change by separating strategic customers from other important customers. This topic will be covered in Chap. 2.
- *Building Block 2: Setting the joint transformation agenda.* In Chap. 3 we focus upon the joint discovery of value innovation opportunities.
- *Building Block 3: Guiding the business transformation for the customer.* In Chap. 4 we explore the concept of customer business transformation and distinguish eight 'business-altering experiences' to guide business transformation and make customers more successful in their markets.
- *Building Block 4: Enabling the internal transformation within the own organization.* In Chap. 5 we focus upon the internal transformation required to enable business transformation with strategic customers.
- *Building Block 5: Undertaking the transformative journey.* In Chap. 6 we focus on the joint transformative journey of the customer and supplier. This chapter is about building a learning partnership at all touch points in the customer-supplier interaction and about fostering 'thriving relationships'. Additionally it is about challenging the (implicit) assumptions that may hamper change.

Transformational sales requires an integrated and holistic approach. As commented by Piercy (2010, p. 354) 'The challenge to strategic customer management mandates effective approaches to cross-functional integration around value processes'. As argued by a European executive on a SAMA-conference

(Strategic Account Management Association), this requires to converge 'strategic management, change management and process management' into the sales function (Seidenschwartz 2005 as cited in Piercy 2010).

We believe that we need to converge and integrate different perspectives, also conceptual ones. Hence this book builds upon academic research in the field of strategic sales and customer management, strategic purchasing and supplier management and business transformation (leadership and change management).

Combined with the insights, experience and best practices from numerous executives in strategic sales and strategic purchasing practice, we offer a framework for transformative buyer-seller relations. In case you would like to add your own insights and best practices, please let us know! (dingena@mpcn.nl and waldemar@pfoertsch.com). Also any other comments or suggestions are more than welcome.

Above all we hope this book inspires you to make a difference and guide life-changing and company-altering transformations.

References

Adamson, B., Dixon, M., & Toman, N. (2013). Dismantling the Sales Machine. *Harvard Business Review, 91*(11), 103–109.

Anderson, J. C., Narus, J. A., & van Rossum, W. (2006). Customer Value Propositions in Business Markets. *Harvard Business Review, 84*(3), 90–99.

Anderson, J. C., Kumar, N., & Narus, J. A. (2008). Certified Value Sellers. *Business Strategy Review, 19*(1), 48–53.

Ateş, M. A. (2014). *Purchasing and Supply Management at the Purchase Category Level: strategy, structure and performance*. Rotterdam: Erasmus Research Institute of Management (ERIM).

Cordón, C., & Vollmann, T. E. (2008). *The Power of Two. How Smart Companies Create Win-Win Customer-Supplier Partnerships That Outperform the Competition*. Basingstoke: Palgrave MacMillan.

Day, G. S. (2011). Closing the Marketing Capabilities Gap. *Journal of Marketing, 74*(4), 183–195.

Dixon, M., & Adamson, B. (2011). *The Challenger Sale. Taking Control of the Customer Conversation*. New York: Portfolio, Penguin.

Gupta, S., & Lehman, D. R. (2005). *Managing Customers as Investments*. Upper Saddle River: Wharton School Publishing.

Ingram, T. N., LaForge, R. W., & Leigh, T. W. (2002). Selling in the new millennium: a joint agenda. *Industrial Marketing Management, 31*, 559–567.

Kotler, P., Kartajaya, H., & Setiawan, I. (2010). *Marketing 3.0.* Hoboken: John Wiley & Sons.

Moss Kanter, R. (2008). Transforming Giants. What kind of company makes it its business to make the world a better place? *Harvard Business Review, 86*(1), 43–52.

Piercy, N. F. (2010). Evolution of Strategic Sales Organizations in Business-to-Business Marketing. *The Journal of Business & Industrial Marketing, 25*(5), 349–359.

Piercy, N. F., & Lane, N. (2009). *Strategic Customer Management.* Oxford: Oxford University Press.

Pine, B. J., & Gilmore, J. H. (2014). A Leader's Guide to Innovation in the Experience Economy. *Strategy & Leadership, 42*(1), 24–29.

Quinn, R. E. (1988). *Beyond Rational Management: Mastering the Pardoxes and Competing Demands of High Performance.* San Francisco: Jossey-Bass.

Quinn, R. E. (1996). *Deep Change. Discovering the Leader within.* San Francisco: John Wiley & Sons.

Quinn, R. E. (2004). *Building The Bridge as you walk on it.* San Francisco: John Wiley & Sons.

Rackham, N., & De Vincentis, J. R. (1999). *Rethinking the Sales Force. Redefining Selling to Create and Capture Customer Value.* New York: McGraw-Hill.

Senn, C., Thoma, A., & Yip, G. S. (2013). Customer-Centric Leadership: How to Manage Strategic Customers as Assets in B2B Markets. *California Management Review, 55*(3), 27–59.

Sherman, S., Sperry, J., & Reese, S. (2003). *The Seven Keys to Managing Strategic Accounts.* New York: McGraw-Hill.

Speakman, J. I. F., & Ryals, L. (2012). Key Account Management: The Inside Selling Job. *Journal of Business & Industrial Marketing, 27*(5), 360–369.

Tarasi, C. O., Bolton, R. N., Hutt, M. D., & Walker, B. A. (2011). Balancing Risk and Return in a Customer Portfolio. *Journal of Marketing, 75*(3), 1–17.

Vitale, R., Giglierano, J., & Pfoertsch, W. (2011). *Business-to-Business Marketing: Analysis and Practice.* Boston: Pearson Education.

Wießmeier, G. F. L., Thoma, A., & Senn, C. (2012). Leveraging Synergies between R&D and Key Account Management to Drive Value Creation. *Research-Technology Management, 55*(3), 15–22.

Yip, G. S., & Bink, A. J. M. (2007a). Managing Global Accounts. *Harvard Business Review, 85*(9), 103–111.

Yip, G. S., & Bink, A. J. M. (2007b). *Managing Global Customers.* Oxford: Oxford University Press.

Zander, R. S., & Zander, B. (2000). *The Art of Possibility.* New York: Penguin Books.

2
Driving Change with Strategic Customers

> *The biggest mistake a company can make is to select [strategic] customers solely on the basis of its current sales to those customers*
>
> Yip and Bink (2007a, p. 106)

Successful companies such as ABB, IBM, Bombardier, GE, HP, P&G, SAP, and Vodafone have proven that the value of the customer base is a strong indicator for company value. In this respect several authors have argued that strategic customers should be managed as assets and therefore time and money spent should be regarded as investments rather than expenses (e. g. Ingram et al. 2002; Gupta and Lehman 2005; Tarasi et al. 2011; Senn et al. 2013). Time and resources invested in acquiring customers and maintaining customer relationships should reflect the value customers are expected to generate over time. By reducing vulnerability andrisk in the customer portfolio and increasing the focus on (potential) valuable customers, company value can be enhanced. Hence managing customer portfolios effectively and differentiating strategic customers from others is vital to company success.

Customers are not equal. By analyzing our customer portfolio we are able to describe the mix and value of different customers and prospects we currently have. One reason for doing this is to understand how balanced and healthy our customer portfolio is and to assess the business risks (Piercy and Lane 2009). Topics to explore include for example the dependency on large customers and the focus on future potential. Another reason is to differentiate between customers in such a way that we are able to match different sales approaches and allocate resources effectively. As argued by Ingram et al. (2002, p. 561) 'the type of relationship and the selling model used for each customer group must balance customer value and cost'. In this book we focus upon building transformational relationships with strategic customers. We believe that joining forces with strategic customers can be a driving engine to realize change and make a difference in the marketplace. Prerequisite is that we understand who these strategic customers really are.

Separating Strategic Customers from Others

The way in which strategic customers are distinguished from others differs from one organization to another. In practice we quite often find pyramids with customer tiers or 'A/B/C' rankings. Also the definition of strategic customers varies and criteria are company specific. Terho (2009) provides an extensive literature overview of academic portfolio models including dimensions for customer analysis. From a practitionersperspective also the SAMA has published several overviews of criteria used in practice (see also Dingena 2002; Sherman et al. 2003; Yip and Bink 2007). By combining the literature and practitioners' perspectives we propose to take at least two dimensions into account when classifying customers and distinguishing strategic customers from others:

- First by differentiating between customers based upon their **current and potential value to us** (see Sect. 2.1).
- Second by differentiating between customers based upon **our (potential) value to them**. Even though some customers may seem valuable and strategic to us, this does not always mean there is room to build a strategic business partnership. To assess the potential for building balanced strategic business relations, both an understanding of the importance and level of sophistication of purchasing and a correct estimation of our value to the customer is required (see Sect. 2.2).

Based on the abovementioned dimensions we would like to define a strategic customer as follows:

> A **Strategic Customer** is: a customer whose current and potential *value to us* is high and *to whom our (potential) value* is significant as well. It is a customer who makes us *change* and who is willing and committed to change with us.

In appendix A, we provide a decision logic to guide you in separating your true strategic customers from others. Once we have distinguished true strategic customers from the other customers, we have a solid basis to differentiate sales approaches and resource allocation accordingly. With our strategic customers we can drive change and transform the business (see Sect. 2.3).

2.1 Value of Customers: Do They Make us Change?

To assess the current and expected value of customers, it is important to take two notions into account:

- The past or current value does not predict future value.
- Value should be measured beyond financial metrics.

Current Value Does not Predict Future Value

Many companies use current size of the customer in terms of sales volumes, revenues or profit as one of the dimensions to determine the importance of customers. Current size indeed matters and can be used as a starting point (80/20-rule). However it should not be the only decisive factor to rely upon. Large customers are not necessarily strategic customers (e. g. Rackham and De Vincentis 1999; Yip and Bink 2007a). Next to current value (looking backwards), it is important to look at future or potential value of the customer. To assess future value first of all requires making the time horizon of the customer assessment explicit (e. g. a 2–3 year timeframe). It also requires a selection of indicators to be used to assess future potential. As argued by Taleb (2007), we cannot predict the future based upon our past experience. Results in the past are not a guarantee for future results. As referred to in the sales literature: 'When your headlights aren't on, the best rearview mirror available isn't likely to improve your driving' (e. g. Peppers and Rogers 2012, p. 63). To see potential future value it is important to look forwards and prevent ourselves from a 'myopic' view blurred by past experience.

So how can we see our customers' future potential? Indicators can be found by looking through three lenses (see Fig. 2.1): *growth within* the customer, *change* in relevant areas and *growth of* the customer.

> First there is the potential to increase 'our share of the cake' within the customer. Through this lens we explore our potential to increase our 'share of wallet' or other unrealized potential value within unexplored areas, departments or functions of the customer.
>
> The second more revealing lens is to look at the way the customer is changing in relevant areas, for example in their strategies to increase investments or expenditures in particular (synergistic) markets or technologies.
>
> A third important lens to assess is the total growth of the customer ('increase of the cake').

3 Lenses	Indicators to assess Potential Value
Growth *within* Customer	• Current 'share of wallet' • Unexplored areas, departments, locations within the customer organization • Unexplored functions and activities within the 'customer activity cycle'
Change in relevant areas	• Strategy to increase investments or expenditures in relevant areas or markets • Strategic change in relevant area, market or region/ strategic fit • Technological developments or change (technological fit or synergy) • Increase of strategic purchasing orientation
Growth *Of* Customer	• Technological, innovative or market leadership ('Best in Class') • Highest growth in their sector • Active in growing or promising markets, regions, technologies • Learning or innovative culture • Financial stability and solvency

Fig. 2.1 Examples of indicators to assess potential value of customers

To which extent are customers for example 'Best in Class' in terms of market or technology leadership? Are they active in growing or promising markets, regions or technologies? Are they learning, growing, competitive and innovative? To which extent are they (financially) stable and trustworthy in the longer run? In Fig. 2.1 examples of input criteria that are used in practice to assess customers' future potential are summarized. This list is not meant to be exhaustive, but rather to be inspirational and to start discussions on what might be relevant criteria in your own and customer's business.

Value Beyond Financial Metrics

To be able to assess a customer's current and future value it is important to determine upfront what is of value to us. Starting point in many cases is the customer's financial or monetary value in terms of volume, revenues, costs to serve and profitability. It is also worthwhile to broaden the perspective to nonfinancial metrics. Examples are relationship value and strategic value.

Relationship Value

Relationship value reflects the strength and quality of the relationship with a particular customer. Peppers and Rogers (2011) describe seven characteristics of a genuine business relationship (see Exhibit 2.1). To which extent would we say the relationship meets these characteristics and has evolved and resulted in a level of mutual trust?

> **Characteristics of Genuine Business Relationships**
> - Mutual
> - Interactive
> - Iterative
> - Benefit for both parties
> - Change in behavior for both parties
> - Unique
> - Trust
>
> Source: Peppers and Rogers (2011, p. 41).

Exhibit 2.1 Characteristics of genuine business relationships

In general we observe that the better the relationship, the higher the chance that the relationship with a particular customer gives us *access* to specific knowledge or know-how, technologies, markets, networks or other customers. Senn (2012) provides a quick test to audit the quality of thecustomer relationship and to determine the future potential of collaboration between the seller and buyer organization. Indicators on a strategic, functional and organizational level may be used to assess the quality of the relationship. Rating the statements in Exhibit 2.2 may give you a first impression.

A strong relationship may not only lead to future business with the customer, it may also lead to increased value in other, indirect ways (e. g. Verbeke and Nagy 2000; Ritter and Gemünden 2003)

- *Referral value*: the extent to which a customer is able and willing to 'refer' us to other potential customers or business. In practice the 'referred' value may far exceed the financial value of the customer itself (e. g. Kumar 2007).
- *Network value*: the extent to which a customer is able and willing to give us access to knowledge, people and/or valuable resources within their network.
- *Reputation value*: the reputation effect of doing business with a customer in the market.

In addition to the relationship value of the customer it is worthwhile to assess their strategic value.

> **Assessing the Customer Relationship**
>
> *Statements about the relationship on a strategic level*
>
> - We have developed a three-year vision of what we want to accomplish with this customer
> - With this customer, we have established strong and personal multi-level contacts
> - We use all our knowledge and skills to generate new and profitable business for this customer
>
> *Statements about the relationship on a functional level*
>
> - Our tailored selling approach is fully supported by the customer
> - We have sufficiently aligned our processes with those of this customer
> - Our systems enable us to demonstrate the total value we deliver to our customers
>
> *Statements about the relationship on an organizational level*
>
> - Our structure promotes cross-functional cooperation with this customer
> - The responsible people for this customer are perceived as credible and trusted advisors by the customer
> - The responsible people for this customer have the skills to generate win-win business opportunities
> - The people in charge of this customer have sufficient authority/decision-making competence to drive the customer business proactively
>
> Based on Senn (2012, p. 38)

Exhibit 2.2 Assessing the customer relationship. (Based on Senn (2012, p. 38), reproduced with permission)

Strategic Value

Strategic value reflects the extent to which a particular customer is helping (and sometimes forcing) us to move into our chosen strategic direction. This may be a strategic focus on particular markets, competencies or technologies, or plannedinvestments in technological or market innovation. Does the direction in which the customer is moving suit our strategic direction (*strategic fit*)? Is this a customer who 'makes us change'? Strategic customers are typically customers who make us change. In a close cooperation between suppliers and customers joint value innovation takes place, in many cases transforming the seller and buyer business. See Exhibit 2.3 for an example of joint innovation with strategic customers in the automotive industry.

> **Joint Innovation with Strategic Automotive Customers at Kendrion**
>
> Kendrion develops, manufactures and markets innovative high quality electromagnetic systems and components for their automotive and industrial customers. One of the business units, *Kendrion Passenger Car Systems* is a globally well-known partner within the automotive industry. Joint innovation takes place together with strategic automotive customers.
>
> Dr. Bernd Gundelsweiler (CEO Division Automotive, Kendrion): 'The automotive market is changing rapidly. Environmental protection and sustainable mobility are key issues. Reduction of fuel consumption and CO_2 e/NOx emissions are our customers' top priority. Through joint innovation with our strategic customers we contribute to increased fuel efficiency and in the longer run to emission free driving. In order to comply with European environmental regulations, one of our strategic Customers – a leading automobile manufacturer – is working closely together with their suppliers to improve exhaust values. In order to support the thermal management of engines we jointly developed a valve through simultaneous engineering.
>
> The joint innovation process, starting with the definition of functions to be improved up to the fully automatic production line of engine valves, took around 2.5 years. During this process, working closely together in cross company project teams, the weight and the size of the valve were reduced by more than 50 percent. The joint innovation resulted in a significant improvement and a high cost reduction on a yearly basis for our customer. This joint innovation is a true win-win for both parties. Next to the mentioned benefits for our customer and the environmental contribution, the co-created value resulted in an additional yearly turnover for Kendrion of several million Euro, contributing to our leading position in the global market of engine valves'.
>
> Joint innovation with strategic customers is part of a broader innovation program at Kendrion. Piet Veenema, (CEO Kendrion): 'By continually investing in innovative processes, R&D, testing facilities and high-end production units Kendrion is always ready to supply its customers worldwide with the high-tech solutions they need'.
>
> Source: Interviews with Dr. Bernd Gundelsweiler (CEO Division Automotive, Kendrion) and Piet Veenema (CEO Kendrion), September 2013.

Exhibit 2.3 Joint Innovation with strategic automotive customers at Kendrion. (Based on interviews with Dr. Bernd Gundelsweiler (CEO Division Automotive) and Piet Veenema (CEO Kendrion) in September 2013, published with permission)

Based on their current and potential value we can distinguish four types of customers (see Fig. 2.2): Transactional, Development, Large and potentially Strategic customers.

Transactional customers are customers with low current value and low expected growth potential (expected future value). They may become valuable if we succeed to streamline our sales approach accordingly (see Sect. 2.3).

Development customers are those customers and prospects who may have a low actual value today, who however have a high growth potential (expected value) in the future. Because of their current low value, chances are that in practice development customers do not get the attention they should get.

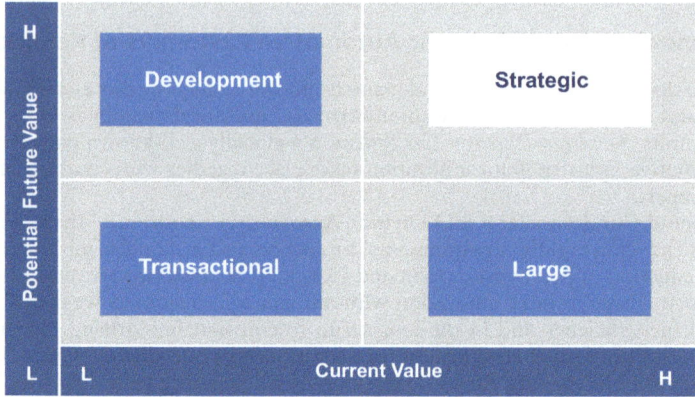

Fig. 2.2 Separating strategic customers from others

Large (or currently valuable) customers have a high current value but a low growth potential. In practice large customers may get more time and resources than can be justified. It takes courage to break out of the comfort zone of habitual routines and established relations that are built over time in these customer-supplier interactions.

Potentially Strategic customers are customers with a high current and high potential value. True strategic customers can be differentiated from other important customers if they are equally willing to invest in a business partnership with our company as well (see Fig. 2.3). Before we decide to distinguish a customer as a strategic customer it is not only important to assess if they are important to us, but how important we are to them.

To assess the potential for building balanced strategic business relations, an estimation of our value to the customer is required.

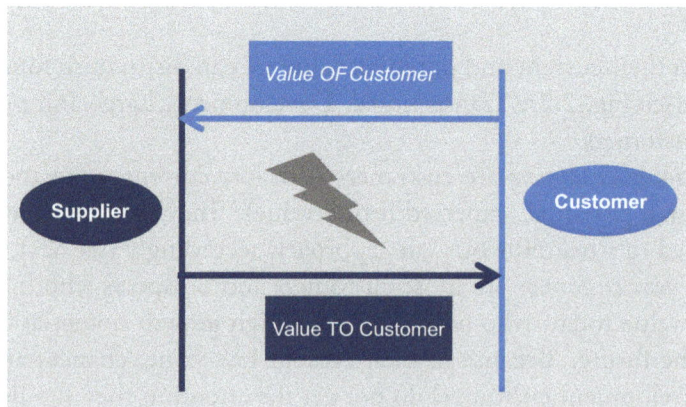

Fig. 2.3 Match or mismatch?

2.2 Value to the Customer: Are They Willing and Committed to Change with us?

Even though some customers may seem valuable and strategic to us, this does not always mean there is room to build a strategic or transformative relationship. As Gosselin and Heene (2003) suggest, strategic customer-supplier relations are relations that are of strategic importance both to the supplier and to the customer. In practice it will be possible to build transformative relations with those strategic customers who are interested and committed to change with us. To assess the potential for building balanced business relations and customer's willingness to change with us, three indicators may be used:

- The importance of purchasing within the customer organization.
- The level of purchasing sophistication or maturity within the customer organization.
- The customer perspective upon our importance in their purchasing strategy and purchasing portfolio.

Importance of Purchasing within the Customer Organization

A first indicator for assessing whether we might be able to build strategic business relations with customers is to assess the importance of purchasing in general within the customer organization. In his foundational article: 'Purchasing must become Supply Management,' Kraljic (1983) argues that a customer's need for a supply strategy depends upon two factors:

- First, the strategic importance of purchasing in terms of financial impact within the customer organization;
- Second, the complexity of the supply market.

One way to assess financial impact is to look at the 'purchasing value in relation to the costs of goods sold' (van Weele 2014, p. 12). In practice this purchasing value is also referred to as a 'purchasing ratio'. According to van Weele (2014), the average purchasing value is around 50 percent. Data from an International Purchasing Survey project (Wynstra 2009) shows that this average purchasing ratio varies between industries, for example, around 40 percent in financial and professional services firms up to almost 60 percent in vehicle manufacturing (see Fig. 2.4).

The financial impact of purchasing within the organization is leveraged by the height of the purchasing ratio. As demonstrated by the so called 'Du Pont

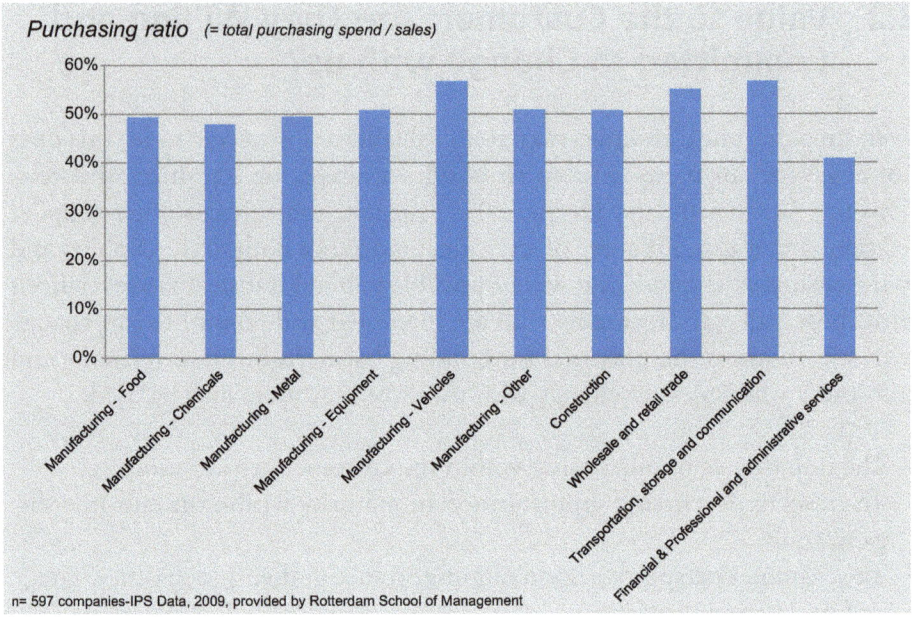

Fig. 2.4 International Purchasing Survey: Purchasing Ratios across industries. (IPS Data 2009, provided by Finn Wynstra, Rotterdam School of Management, reproduced with permission)

analysis', the higher the purchasing ratio, the higher the financial impact of (cost savings through) purchasing.

As illustrated in Fig. 2.5, at Heineken a cost saving in purchasing of 2 percent results in an improvement of the return on capital employed of 9.4 percent. As commented by van Weele (2014), the financial leverage at Heineken is rather low as a result of the relatively low capital turnover ratio. In companies with a higher velocity (higher capital turnover) a much higher leverage of purchasing savings on the company's return on investment can be found.

In several industries (such as automotive manufacturing and Information and Communication technologies (ICT) we find purchasing ratios up to 75 percent, indicating a large impact of purchasing on the company's profitability. According to Axelsson et al. (2005) the sourcing transformation at IBM for example resulted in a ratio of 72 percent of the costs of goods sold (see Exhibit 2.4).

In general we can say the higher the importance or strategic relevance of purchasing in a customer organization, the higher the chance that customers are developing strategic sourcing strategies and developing strategic relations with their key suppliers.

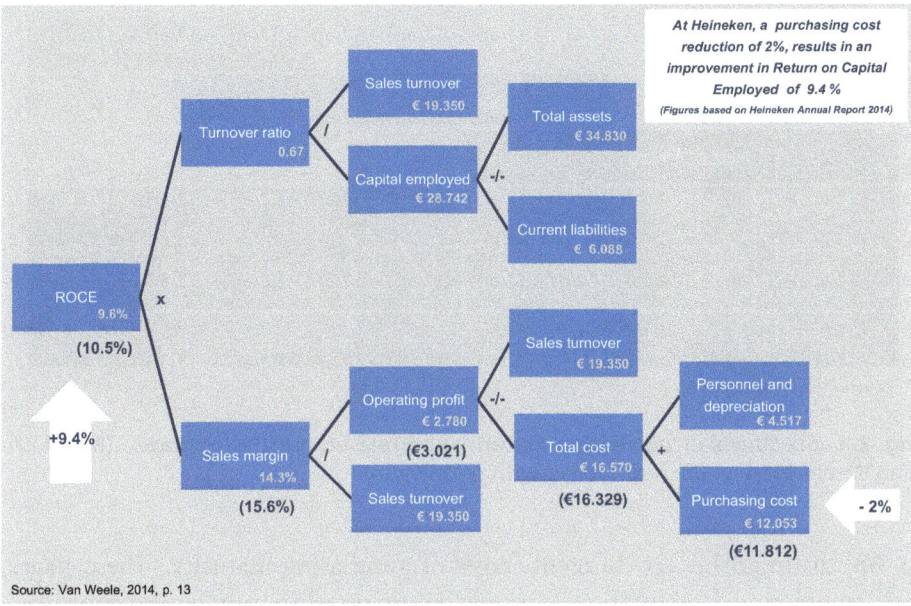

Fig. 2.5 Dupont Analysis Heineken NV (2014): Impact of purchasing savings on Return on Capital Employed. (Van Weele 2014, p. 13, updated by Van Weele with data 2014 in April 2015, reproduced with permission)

Sourcing Transformation at IBM

'IBM's sourcing transformation story begins in the early 1990's. IBM suffered from shrinking earnings and cash flows. In 1993, earnings and cash flow went negative and the stock price decreased to US$ 41, while it was US$ 117 in 1990. The fate of the entire company was at stake. This 'near-death' experience' acted as a catalyst for change. IBM embarked on a transformation of all its key business processes, resulting in many role changes. Sourcing or 'procurement' as it is usually referred to within IBM, was recognized as a key part of the overall IBM corporate transformation, sponsored by the CFO and CEO. Procurement was becoming increasingly important as IBM's business model changed from being a hardware manufacturer towards being a service provider. Today almost 50 percent of each revenue dollar is converted to supplier spending and purchasing makes up 72 percent of the costs of goods sold.'

Source: Axelsson et al. (2005, p. 7–8).

Exhibit 2.4 Sourcing transformation at IBM

Level of Purchasing Sophistication or Maturity

A second indicator for assessing whether we might be able to build strategic business relations with customers is to assess the level of purchasing sophistication or maturity within the customer organization. Van Weele (2014,

			'Total Customer Value'		
		'Total Cost of Ownership'			
'Best Price'					
Stage 1	Stage 2	Stage 3	Stage 4	Stage 5	Stage 6
Transactional Orientation	Commercial Orientation	Purchasing Coordination	Internal Integration	External Integration	Value Chain Integration
Purchasing as a functional approach			Purchasing as a crossfunctional approach		

Based on Van Weele 2014, p. 68

Fig. 2.6 Six stages of purchasing maturity and related purchasing focus. (Based on Van Weele 2014, p. 68, reproduced with permission)

p. 68) distinguishes six different stages of purchasing maturity. These stages are: transactional orientation, commercial orientation, purchasing coordination, internal integration, external integration and value chain integration. According to van Weele (2014) these stages can be divided into two different approaches to purchasing. Purchasing as a functional approach (stages 1–3) and purchasing as a cross functional approach (stages 4–6). Depending on the level of purchasing professionalism the focus of the purchasers involved may vary from a primary focus on getting the 'best price' in the earlier stages, towards a focus on 'total cost of ownership' and 'total customer value' in the later stages. See Fig. 2.6 for the six stages of purchasing professionalism and the related purchasing focus. The different stages are elaborated in more detail in the following section.

Stages 1–3: Purchasing as a Functional Approach

In the first three levels of purchasing maturity, purchasing is regarded as a separate function within the buying organization. The lowest level of purchasing sophistication is the *'transactional orientation'*. Purchasing acts as a separate unit or department. This maturity stage can be recognized by a 'very passive or reactive purchasing operation where the purchasing professionals in principle merely 'administer' the purchasing tasks' (Axelsson et al. 2005, p. 21). In this case purchasing professionals are not involved in specifications or supplier selection, but basically 'order' products and services and administer the purchasing. In the *'commercial orientation'* phase the focus on price reduction increases. Purchasing is involved in tenders, comparing various offers from suppliers, and plays a role in (price) negotiations. In the third stage, *'purchasing coordination'* the focus of attention shifts to control over 'purchased volumes, the number of suppliers and purchased items'. This enables the buying company to 'carry out more powerful and coordinated actions – across factories, business units and divisions' (Axelsson et al. 2005, p. 22). The level of coordination increases in stages 2 and 3 and may even lead to a centralization of the purchasing in stage 3. Major focus of purchasers acting on this level of professionalism is to get a 'best price' (in particular in the first two stages) and to start focusing on 'total cost of ownership' in the purchasing coordination stage.

Stages 4–6: Purchasing as a Cross Functional Approach

Moving up to a higher maturity level in purchasing requires a change from a functional to a so-called cross functional and center-led approach. Stage 4, *internal integration,* 'implies that the organization handles purchases and suppliers in a more process-oriented way, utilizing cross-functional teams with the relevant competencies and expertise who naturally take responsibility for important goods and services (functions) bought (…). Such organizations utilize not only pre-qualified but also ranked suppliers that are in various ways connected to development and improvement programs supported by performance-based contracts' (Axelsson et al. 2005, p. 22). In this stage relevant competencies are bundled for each purchased category, i. e. a (virtual) team is composed with specialists from different areas influencing the problem definition, specification and supplier selection. The decision making unit is deliberately extended in such a way that all specialists (including users) contribute to the purchasing process. Stage 5, *external integration*, includes the synchronization and optimization of the upstream supply chains. This requires not only close collaboration with all 'internal customers' but also with suppliers, including supportive (web based) EDI systems and (web-enabled) collaborative planning. Stage 6, *value chain integration* includes not only the integration with suppliers but also with the buying organization's own customers. This includes 'all of the synchronized purchasing and supply operations from the previous phases plus actively contributing to the creation of customer value, for example in the form of superior quality, functionality and availability of final products. In-depth understanding of customer needs and willingness to satisfy them are the basic requirements for reaching phase 6. This presupposes that sourcing, in addition to the demands of the previous steps, also has a global perspective on suitable suppliers and is sufficiently positioned and equipped to undertake entrepreneurial collaboration with other suppliers' (Axelsson et al. 2005, p. 22). In this stages the focus of purchasing moves beyond total cost of ownership, towards the total customer value (both in terms of cost reductions and revenue streams).

Leading Edge Companies Redefine Buying and Selling Roles

Depending on the level of purchasing maturity, purchasing may contribute to a company's competitive advantage in various ways. Wynstra (2013; see also Axelsson et al. 2005) distinguishes three performance areas where purchasing may have an impact on the company performance:

- Cost Management: this may include a lower purchase price, lower transaction costs, lower overhead costs, higher cost productivity and increased logistics efficiency.
- Asset utilization: for example capital utilization, cash velocity, inventory management, cycle time reduction and payment terms.
- Revenue growth: as a result of for example new products or services, first to market, flexibility and responsiveness and improved quality.

Obviously the customer focus on these performance areas correlates with the level of purchasing maturity. Starting with lowering purchasing prices at the stage of commercial orientation, via total cost of ownership and improved asset utilization in the stages of coordination and integration towards total customer value in the stages of external and value chain integration. According to Axelsson et al. (2005, p. 16), 'Purchasing has traditionally focused on cost optimization' whereas 'value creation' is for most organizations, the newest area in terms of sourcing strategy development'. Based upon a global survey of chief procurement officers (CPO's), Hardt et al. (2007) come to a comparable conclusion (see Exhibit 2.5).

In recent academic research on purchasing and supply management, Ateş (2014, p. 13) also concludes that over 'the past two decades, purchasing has evolved from a clerical function of buying goods and services at a minimum price into a strategic function focused on value creation and achieving competitive advantage'.

It is worthwhile to assess the level of purchasing maturity of strategic customers for two reasons: First to start conversations at the right level of consciousness; second to understand whether the customer would be ready to and interested in building transformative relationships with its suppliers. Leading edge companies understand the power of transformational buying and selling relationships with selected strategic partners. Also Cordón and Vollmann (2008) elaborate various examples both on the buying side (for example Honda) and on the selling side (for example ABB) of what they refer to as 'super-collaborative' relationships (see Exhibit 2.6).

In a recent workshop including 30 academics in the Purchasing field the question was raised which are the most important future hot topics in Pur-

Inventing the 21st Century Purchasing Organization (McKinsey Survey of > 200 CPO's)

'Our survey [...] suggests that the role of purchasing in many companies hasn't evolved much beyond the function's narrow transactional roots [...] But some purchasing and supply-management organizations are attracting the attention of CEOs by taking the function to the next level. By integrating their activities more closely with those of their internal customers, some purchasing units have gained sustainable cost reductions in nontraditional areas (marketing or health benefits for example) where previous optimization efforts have fizzled. Others go further still, using their insights to challenge and enhance manufacturing or administration activities. Finally, some [high performers] use purchasing as a springboard to innovation, leveraging a broader supply base of tangible and intangible goods to enhance product-development efforts. Top-performing companies redefine their [the purchasing] role and ensure that its goals align with corporate strategy'.

Source: Hardt et al. (2007, p. 1)

Exhibit 2.5 Inventing the 21st century purchasing organization (McKinsey survey of > 200 CPO's)

Strategic Collaboration with Suppliers at Bombardier Transportation

Bombardier Transportation ($ 8.8 billion annual sales as of end 2013) is the rail division of Bombardier Inc (global manufacturer of planes and trains). Bombardier Transportation is a world leader in the design, manufacturing and support of rail equipment and systems. 'Bombardier Transportation (BT) recently created collaborative relationships with ten strategic suppliers. In each case, significant changes were required in both BT and the supplier. Included are new ways to share information, joint innovation efforts, product standardization, modularity, process simplification, and new approaches to negotiation – all supported by increased trust. The results have been very important – both for BT and for each of the suppliers. BT started this effort by reducing its supplier base from several thousand to 500. This group was further segmented into two fundamentally different groups: those with which BT can develop collaborative relations, and those that will always be arm's length. The former group (of about 100) was further segmented into about 20 'strategic suppliers' and the rest. The ten collaborative relationships described above come from this set of 20, and the other ten are currently being prepared. In each case, the goal is to create a true partnership – what we call a 'pair of aces'. These 20 pairs of aces are expected to provide roughly 50 percent of the total purchasing expenditure. This implies a major increase in sales volume with each of these suppliers, significant reductions in joint costs of doing business, and (even more important) critical support for new product development. If each of these partnerships can become a pair of aces, BT will have a definitive competitive advantage'

Source: Bombardier (2013) and Cordón and Vollmann (2008, p. 9–10).

Exhibit 2.6 Strategic collaboration with suppliers at Bombardier Transportation

> **Top 10 Future Hot Topics in Purchasing and Supply Management**
> - Innovation: using suppliers as a source of competitive advantage
> - Developing collaborative buyer-seller relationships
> - Aligning the sourcing strategy with the business strategy
> - Engaging internal stakeholders in sourcing activities
> - Human Resources Development in Purchasing & Supply Management
> - Reverse marketing: how to be become a customer of choice for your key suppliers
> - Corporate Social Responsibility and the impact on the value chain
> - Value chain management
> - Supply risk management
> - Value based sourcing
>
> Source: Rozemeijer et al. (2012, p. 64; n = 30 Academics)

Exhibit 2.7 Top 10 future hot topics in Purchasing and Supply Management

chasing and Supply Management (Rozemeijer et al. 2012). The top 10 items are listed in Exhibit 2.7.

Chances to build a strategic business partnership are higher with customers whose maturity in purchasing is higher. Customers who focus on total cost of ownership or preferably on total customer value are more mature than customers who have a transactional orientation and focus primarily on (reducing) prices. Even though all customers talk about price, it is important to look at other indicators to assess their purchasing maturity. One of the interesting topics on the agenda of purchasing professionals is called '*reverse marketing*', which means understanding how to become a 'customer of choice' to the key suppliers (see for example Sony in Exhibit 2.8).

> **Reverse Marketing at Sony**
>
> 'The Internet site of Sony provides an interesting example of how large manufacturers communicate with their suppliers nowadays. New suppliers are explicitly invited to study Sony's purchasing policy, structure and pre-qualification procedure. In fact, Sony uses the Internet as a marketing instrument to inform suppliers about their current procurement vision and strategy. Through this website future 'partners' (as Sony likes to refer to suppliers who are capable of meeting all Sony's requirements) can subscribe and inform themselves about what it takes to become a prospective Sony supplier'.
>
> Source: van Weele (2014, p. 65).

Exhibit 2.8 Reverse marketing at Sony

In order to fully understand the potential to build a balanced relationship with our customers we need to focus on the customer and engage in what we can refer to as *'reverse purchasing'* (as elaborated in the following section).

Customer Perspective: Reverse Purchasing

A third indicator for assessing whether we might be able to build strategic business relationships with customers is to understand our position in their purchasing strategy and purchasing portfolio. This involves looking at ourselves from theirperspective and analyzing how important we may be today or could be in the future (in other words to apply 'reverse purchasing'). From a customer perspective, many companies nowadays carry out a kind of 'purchasing portfolio analysis' as a basis for a differentiated purchasing strategy. Kraljic (1983) proposed an approach to 'shape supply strategies' (Kraljic 1983, p. 112). This approach has been widely adopted and adapted by purchasing professionals and academics. According to Ateş (2014) the Kraljic model has been cited in academic practice over 250 times. In order to develop a product portfolio to classify the purchasing assortment, Kraljic (1983) classifies purchased goods and services into four 'sourcing categories', based upon two dimensions: the 'profit impact' of a given supply item to the company (which we refer to as 'purchasing value') and the 'supply risk' (which we refer to as 'purchasing risk').

Purchasing Value can be defined in terms of the volume purchased, percentage of total purchase cost or impact on product quality or business growth. This dimension covers the perceived relative importance of the purchased category as compared to other categories. It is important to note that even though the term profit impact might suggest a focus on financial aspects only, and purchasers in practice sometimes limit themselves to this dimension exclusively, qualitative elements (e. g. impact on product quality) are also included in the original Kraljc model.

Purchasing Risk can be assessed in terms of availability (short term and long term), number of suppliers, competitive demand, make-or-buy opportunities, storage risks and substitution possibilities. This dimension covers the perceived or experienced risk of the purchased item as compared to other items. How easy is it for a buyer to change to an alternative source of supply? Besides the actual risk (perceived), switching costs may play a role. These switching costs can be financial, costs of time and effort to be spent or relational (Burnham et al. 2003). Relational switching costs involve 'emotional discomfort due to the loss of identity and the breaking of bonds' (Burnham et al. 2003, p. 112).

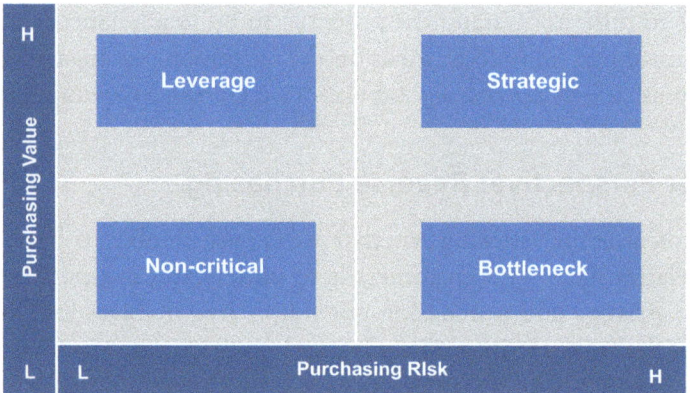

Fig. 2.7 Customer perspective: Kraljic purchasing portfolio

Based upon these criteria, purchasing managers can classify purchased items (and suppliers) into four sourcing categories: Strategic (high purchasing value and high purchasing risk), Bottleneck (low purchasing value and high purchasing risk), Leverage (high purchasing value and low purchasing risk) and Non-critical (low purchasing value and low purchasing risk) (see Fig. 2.7).

The portfolio model and many variations upon the original model have been widely adopted in purchasing practice. Figure 2.8 depicts one example of the application of the model in the context of an automotive company.

The customer using a portfolio model as shown in Fig. 2.7 and 2.8 will most probably build its purchasing strategy accordingly, by building diverse relationships with suppliers based upon the supply situation faced.

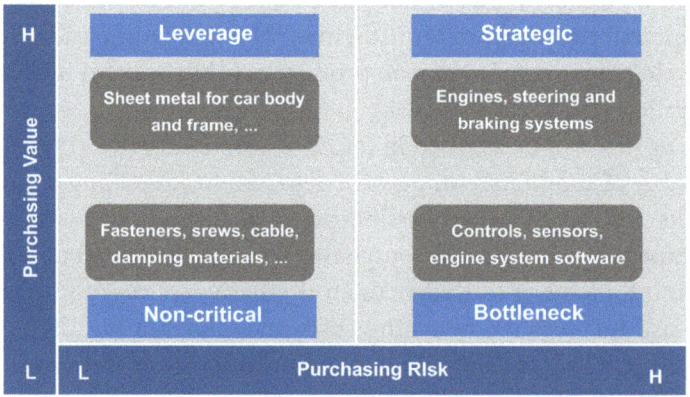

Fig. 2.8 Typical purchasing portfolio for an automotive company

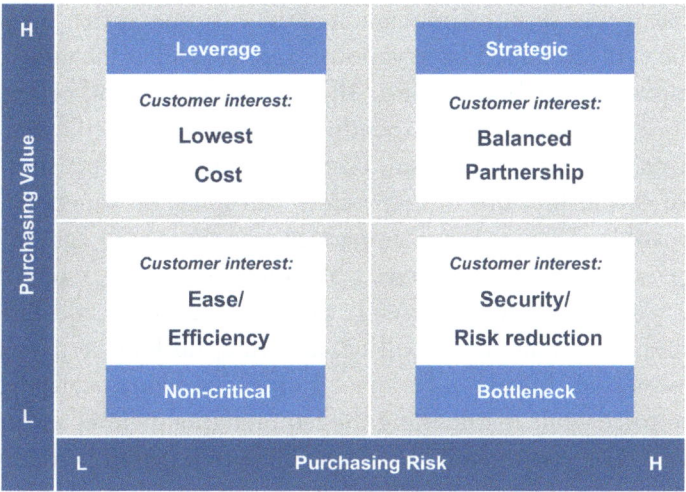

Fig. 2.9 Customer perspective upon supply: reverse purchasing

Depending on our position within the customer purchasing portfolio (as perceived by the customer), a customer may be either explicitly or implicitly looking for a particular type of relationship with a supplier (see Fig. 2.9).

In the *strategic category*, depending on the relative power position of the different parties involved, customers may be striving for a collaborative relationship or 'performance-based partnership' (van Weele 2014, p. 166). Customers may be primarily interested in intensifying relationships with these suppliers for two reasons: First to significantly increase value by joint value/cost improvements, joint innovation, shared new technologies, joint processes etc.; Secondly to minimize supply risk as much as possible. Customers may be looking for strategic business partners with whom they can change the future business and become or remain competitive in their markets. Customers' major concern may be to build up a 'balanced' relationship. The balance of power between buyer and supplier may differ: buyer-dominated, supplier-dominated or balanced (see also Caniëls and Gelderman 2007). In particular in the balanced relationship the purchasing strategy will be aimed at intensifying collaboration and building a partnership. 'The goal is to create mutual participation based on pre-planned and mutually agreed cost and operational improvement targets. A relationship based on 'open costing' is preferred (…). With the suppliers, efficiency programs are developed to achieve cost reduction, quality improvement, process improvement, and improved product development. Such cooperation can in the end lead to fading of borders between the different companies' (van Weele 2014, p. 166).

In the *leverage category,* the impression of the customer is that the purchased items can be easily obtained from various suppliers and at the same time have a strong impact on the bottom line. The major concern and priority of customers will be to reduce cost. Depending on both the specific item and the level of purchasing professionalism, customers will either focus on reducing total cost of ownership, but in many cases will focus on squeezing prices as much as possible. The 'leverage' impact means that small cost savings will have a large impact on money saved. 'A purchasing strategy of competitive bidding or tendering will be pursued. Since the suppliers and products are basically interchangeable, there will be, as a rule, no long-term supply contracts. In most cases, buyers will adopt a multiple sourcing strategy; buying at a minimum price while maintaining the required quality level and continuity of supply will take priority. Small savings (small in terms of percentages) represent a large sum of money. This justifies an active market scanning through continuous market and supply research. Regularly, outsiders will be introduced so as to avoid price arrangements between the present suppliers' (van Weele 2014, p. 166). In many cases purchasing is coordinated and agreements or bulk contracts are centrally (globally) negotiated.

In the *bottleneck category,* customers will try to reduce (perceived) risks and focus 'primarily on securing [the] continuity of supply, if necessary even at additional cost' (van Weele 2014, p. 166). These supplies may have relatively little purchasing value but are regarded by customers as vulnerable in regard to their supply. Perceived risk or uncertainty may stem from different sources. The product or service may, for example, be customized in such a way that identical alternatives are not available on short notice, or the supply of particular ingredients may be influenced by world market fluctuations. Also (perceived) switching costs may be considered as a barrier to explore alternatives. It is important to understand where the major risk is perceived. If customers understand the buying situation as a real bottleneck, activities may be conducted to reduce dependence on suppliers by developing alternative products or suppliers, by insourcing or by logistic solutions. Van Weele (2014) mentions consigned stock agreements, alternative modes of transportation and the active investigation of supply alternatives.

In the *non-critical (referred to by Van Weele as 'routine') category,* few technical or commercial problems can be seen from a purchasing perspective. Both purchasing value and purchasing risk are low. Customers can choose from many suppliers. In purchasing practice many items are within this category. According to van Weele (2014, p. 165): 'Usually, 80 percent of the time and energy of purchasing is used for these products: a reason why purchasing is often seen as an administrative job. The purchasing of these products should be organized efficiently, in order to spare time for the other, more interesting products'. A major challenge for customers in this situation is to reduce administrative and logistic complexity. It is all about finding ways to make it easier to specify, select, order, store, internally distribute, use and pay for these items and services. In order to improve efficiency, customers may standardize the assortment or find ways to subcontract product groups (e. g. office supplies or cleaning and catering services) or improve logistics with web-enabled services (e-catalogue, e-ordering, mobile apps, e-payment). In some cases, (part of) the purchasing of these items is outsourced to specialized purchasing companies.

Potential for Building Strategic Business Partnerships

Obviously chances to build strategic business relationships are highest with customers who perceive or even classify us to be a (potential) strategic supplier. In cases where the balance of power is more or less equal, the relationship may be further developed into a partnership and transformative relation. In some cases, also within the bottleneck situation, we may find possibilities to build strategic relations, depending on the reason why the supply situation is perceived as risky by the customer. Joined forecasting or buffer stocks may be ways to reduce (perceived) risks together with customers. Also pro-active and timely (joint) innovation may prevent customers from searching for alternative products or suppliers. Additionally, the customer's risk perception may be altered (see Sect. 3.1 for a further elaboration of this topic).

In the routine segment, possibilities are low unless we find ways to reinvent the way of buying for the customer and significantly reduce the logistic complexity for the customer. Either by grouping article groups or components that were traditionally purchased from different suppliers (becoming a systems integrator) or by developing solutions that simplify ordering, internal handling/distribution and invoicing.

The potential to build a strategic business partnership with a customer seems to be lowest in the leverage segment. In most cases the purchasing strategy is to obtain the 'best deal for the short term'. Customers will frequently reallocate purchasing volumes over suppliers making use of e-auctions and dynamic pricing and reducing possibilities to develop long term relationships.

For strategic sales teams, 'reverse purchasing' may be helpful to get a clear understanding of where the selling company should be positioned in the buyer's purchasing portfolio. In this way the current and potential value to the customer can be assessed. This may also include aligning perspectives with purchasing teams who may be overly focusing on price reductions rather than cost reductions or potential joint value increases.

Kraljic Revisited

The Kraljic matrix can be a good starting point for understanding our importance and value to our customers. Ateş (2014) argues that to fully understand the 'richness of purchasing strategies' in practice it is important to broaden the perspective and include for example customers' 'competitive priorities' as an additional layer (see Exhibit 2.9).

Potential 'Mismatch' may be Colored by Different Value Perspectives

We indeed believe that adding the customers 'competitive priorities' as suggested by Ateş (2014) is an important additional layer that enriches our understanding of customer purchasing strategies. In this respect it is important to note that part of the potential 'mismatch' (as shown in Fig. 2.3) may be caused by the different mindset or value perspective chosen by sales professionals and purchasing professionals. Value to the customer may be defined in many ways (see Chap. 3 for a detailed discussion on this topic). Lindgreen and Wynstra (2005) observe two 'research streams' in the literature on defining value in business markets. One perspective on defining value is based on the 'value of goods and services', another perspective is based on 'the value of buyer-seller relationships'. We observe these different perspectives to exist in practice as well. Purchasers assessing the value of the (potential) supplier offering based on the value of the goods and services or 'sourcing category' may have a different perspective on the value of the relationship as compared to those who assess value based upon the potential of the 'buyer-seller relationship'. Building transformative relations requires to be able to think beyond sourcing or selling categories and assess the potential value of the relationship. In addition to understand the professionalism of our customer's purchasing strategy and to understand their perspective upon our importance as a supplier (knowing if they apply sourcing models and how we may be plotted), it is also important to understand their value perspective upon the relationship.

> **Building transformative relationships** requires to be able to think beyond sourcing or selling categories and assess the potential value of the customer-supplier relationship

Deepening the Understanding of Purchasing Strategies: include Competitive Priorities

In recent academic research, Ateş states that even though the Kraljic model was introduced more than three decades ago, it is still popular among purchasers in practice and has been cited in academic practice more than 250 times, also inspiring the development of other portfolio models. At the same time such models 'have been criticized because they focus on a limited set of contingencies and are not distinctive enough to fully cover the variety and richness of purchasing strategies' implemented in practice. In her research, Ateş deepens the understanding of purchasing strategies including among others the competitive priorities customers may have. 'Our results suggest that as an additional, complementary layer, differences in competitive priorities must be examined when defining purchasing category strategies' (Ateş 2014, p. 57). Competitive priorities may include cost management, innovation or improved quality, delivery reliability or increased sustainability. Customers may focus their purchasing strategies around their competitive priorities which are grouped by Ateş into: cost optimization, increase of delivery reliability, innovation, 'emphasize all' and 'emphasize nothing'.

Bottleneck: emphasize all
The *emphasize all* strategy is a strategy in which all competitive priorities are emphasized on high levels. Ateş finds that this purchasing strategy is particularly popular in practice in the *bottleneck* situation. 'When the purchase importance is low but supply risk is high, firms cannot afford to focus solely on the cost objective. In such cases, it is important to assure supply and survive the 'lock-in situation'' (Caniëls and Gelderman 2005; van Weele 2000 as cited in Ateş 2014). As articulated by Ateş, since this purchasing strategy 'requires extensive resources and programs,' this strategy is not found to be used often in the routine segment.

Different approach to innovation in the strategic and leverage situation
A clear focus on the competitive priorities innovation and quality are found in the strategic quadrant. Interestingly, this focus is also applied in some cases in the leverage quadrant. According to Ateş, even though customers may require innovation in both situations, the approach may be very different. 'As buyers are more powerful than their suppliers in a leverage situation, they are more likely to demand that their suppliers provide innovation without a great deal of commitment, whereas for more strategic products it would be more beneficial to participate in joint innovation projects'.

As cost saving has been a traditional topic in purchasing strategies, it is used in several categories, except for the bottleneck situation. As mentioned before, customers use a more 'nuanced' purchasing strategy in bottleneck situations (Ateş 2014, p. 56).

Within the non-critical segment, both the 'emphasize nothing' and accurate time delivery (even at higher cost) are found to be popular.

Source: Ateş (2014) and interview with Melek Ateş March 2014.

Exhibit 2.9 Deepening the understanding of purchasing strategies: include competitive priorities. (Source Ateş (2014), and interview with Melek Ateş Mach 2014, reproduced with permission)

Summarizing: Are They Willing and Committed to Change with us?

A deep understanding of our customers' purchasing strategies will help to assess our importance and value to customers and decide whether we find a 'match or mismatch'. To summarize we can assess our (potential) value to our customer by answering three questions:

1. How *important is purchasing* in the customer organization in general. An indicator may be the total purchasing value or 'purchasing ratio'. This is the purchasing spent in relation to costs of goods and services sold. The higher the purchasing ratio, the higher the chances that purchasing is perceived to be of strategic importance within the customer organization and dealt with in a strategic way.
2. How *mature is purchasing* in the customer organization. This may be based on the level of purchasing sophistication. Is the purchasing function primarily focused on lowering prices or rather focused on decreasing total cost of ownership or increasing competitive advantage (total value)? In many cases this is already visible in the way in which purchasing is organized (is there a functional or cross functional approach and is purchasing elevated toboardroom agenda or not?). The more sophisticated purchasing is within the customer organization, the higher the changes are that customers are willing and committed to build strategic relations with selected and prioritized suppliers.
3. How *important is our (potential) offering* within the customer's purchasing portfolio and purchasing strategy. To what extent do customers believe we could substantially contribute to increasing their competitive advantage and success in their markets?

Based on our value to the customer we can distinguish the true strategic customers from other important customers. See Appendix A for a decision logic that can be used to determine how to separate strategic customers from others.

2.3 Transformative Relationships: Driving Change!

Selecting strategic customers is the starting point for building transformative relations. It makes no sense to differentiate between customers if we do not intend to differentiate our sales approach accordingly. Even though this sounds logical, in practice the differentiation between the different sales approaches for different customer types is not always clearly visible. Customer differentiation should be the basis for developing meaningful strategies to invest time

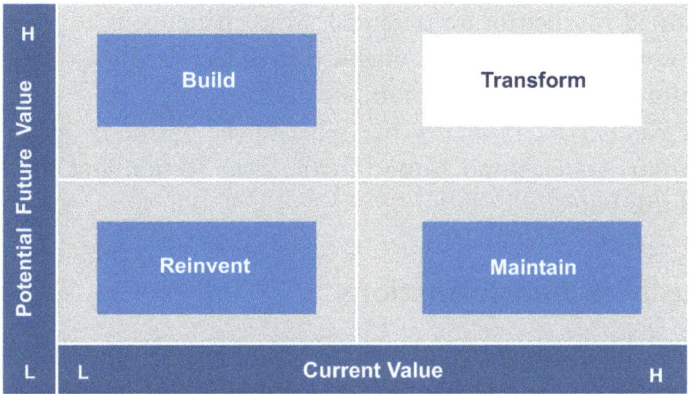

Fig. 2.10 Differentiated customer strategies

and allocate resources in the right way. Unprofitable customers do not exist, however unprofitable sales and service concepts do. In order to clarify and communicate the different approaches towards different customer types, it may be useful to elaborate the major differences in the sales strategy and service level. In addition to the sales strategy and major focus, attention may be paid to the type of interaction (for example the customer-supplier interface, the level, depth and frequency of contacts and the interaction model), the account planning process and the differences in services included per customer type. See for example the template in Appendix B that could be used to differentiate the sales approaches in practice. Even though the elaboration of the differentiated sales strategies towards different customer types goes beyond the scope of this book, the generic customer strategy towards the four customer types is depicted in Fig. 2.10 and is summarized below.

Transactional customers: *Reinvent the sales model* to increase efficiency and reduce costs to serve. The essence of this strategy is to explore possibilities to reinvent the sales and service approach so that customers are served in a more profitable way (for example less face-to-face interaction, increased web enabled sales, increased rationalization).

Development customers: *Build* relations to grow to their fullest potential, by broadening the decision making unit or buying center, establishing relations with new and relevant stakeholders within the customer organization, and building multi-level relations between both organizations in order to be able to release the untapped potential. Social media may be a useful aid in this process. Being connected online to so many will enable us to get in touch with new people in the customer organization easier than before.

Large customers: *Maintain* relations at reasonable cost. Given their size and power, these customers may ask and receive more time and resources than can be justified by their future potential. The challenge is to find a way to maintain the relationship at a reasonable (often decreased) cost and move out of the comfort zone of habitual routines.

Strategic customers: *Transform* the seller-buyer business: guide business transformations (as elaborated upon in the next section and the following chapters of this book).

Guide Business Transformations

The focus of this book is on driving change and making a difference with strategic customers. The essence of these relationships is to guide business transformations. Piercy (2010) underlines that the change in sales focus is moving from a transactional customer approach towards a strategic sales relationship; moving from the 'converting of products and services into cash flow', beyond a focus on 'customer satisfaction and retention' towards 'guiding customers to become more competitive in their markets'.

Also Dixon and Adamson (2011) and Adamson et al. (2012, 2013) propose to move beyond 'solution selling' focusing on buyer needs towards 'insight selling' offering provocative insights into the customer business (see also Nichols et al. 2011). As referred to by Pine and Gilmore (2014), changing the way a customer does business by guiding 'behavior-changing' or 'company-altering' transformative experiences.

Building upon the model of Piercy (2010, p. 354) and including the contributions of Pine and Gilmore (1998, 2014), Dixon and Adamson (2011) and Adamson et al. (2012, 2013), we would like to depict the changing focus from transactional sales, beyond solution sales towards transformational sales as shown in Fig. 2.11.

Building transformative relationships requires not only a 'superior understanding of the customer's own organization, but detailed and insightful knowledge of the customer's end-use markets' (Piercy 2010, p. 353). In addition it requires a profound understanding of available and accessible resources within the own organization and within the value network. This is the starting point for leading change, requiring a 'network perspective' (Senn et al. 2013) of the relationship. 'Through collaboration and matched firm-specific resources and competences, the two firms create a relational value that no one entity could achieve alone; risks are shared and controlled for, too. Network oriented managers thus maintain a continuous stream of insights on the market, the competition, the customer's business, and how the supplier's products, services and capabilities can change the economics of the customer's business model' (Senn et al. 2013, p. 34). See also Prahalad and Krishnan (2008) on the co-creation of value through global networks.

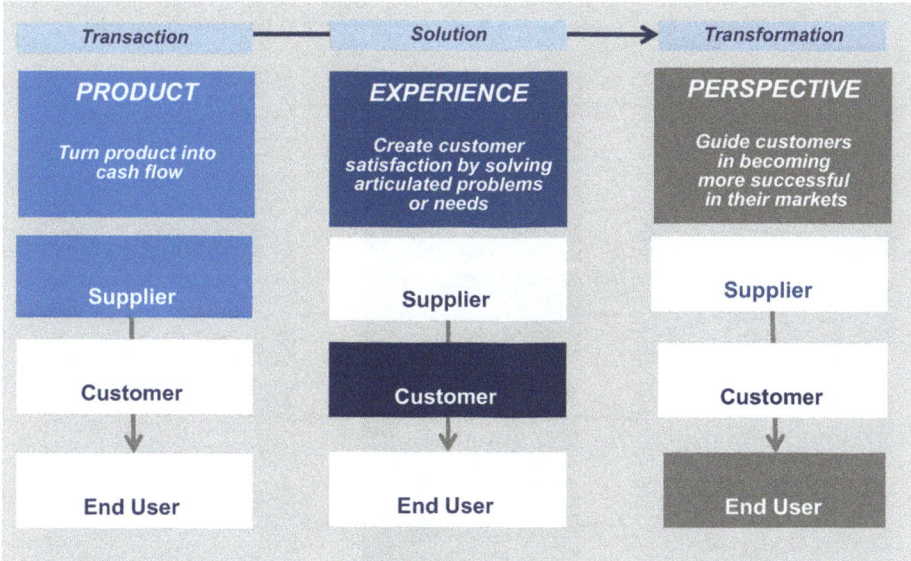

Fig. 2.11 Changing focus: towards transformational sales

Transformational sales is about guiding change and enabling business transformation. It is for both – the customer and the own company. It is about disrupting both the customer's and supplier's thinking and way of doing business. Challenging the explicit and implicit assumptions or mindset salespeople and purchasers may have. 'What is needed is joint transformation; a combined commitment to striking out in a new direction and abandoning the current ways of working' (Cordón and Vollmann 2008, p. 24). Building upon the model presented by Adamson et al. (2013), we distinguish transformational sales from transactional and solution sales as depicted in Fig. 2.12.

We distinguish three value positions a company may achieve: 'be better', 'differentiate' and 'make a difference'. Whereas the supplier purpose in transactional and solution sales approaches may be focused around the first two value positions (being better or being different), transformational sales moves beyond this way of value creation towards 'making a difference' to the customer business (see also Kotler et al. 2010). It moves from demonstrating the value of offerings as compared to alternatives towards providing new perspectives upon customer's business potential and enabling customers to look at their business and their markets in a different way. In order to provide a new perspective upon the way of doing business, suppliers need to move beyond defined needs as expressed by the customer towards the recognition that the customer is in a 'state of uncertainty'. This may be one of the first (implicit) assumptions within the strategic sales team to be challenged: to understand

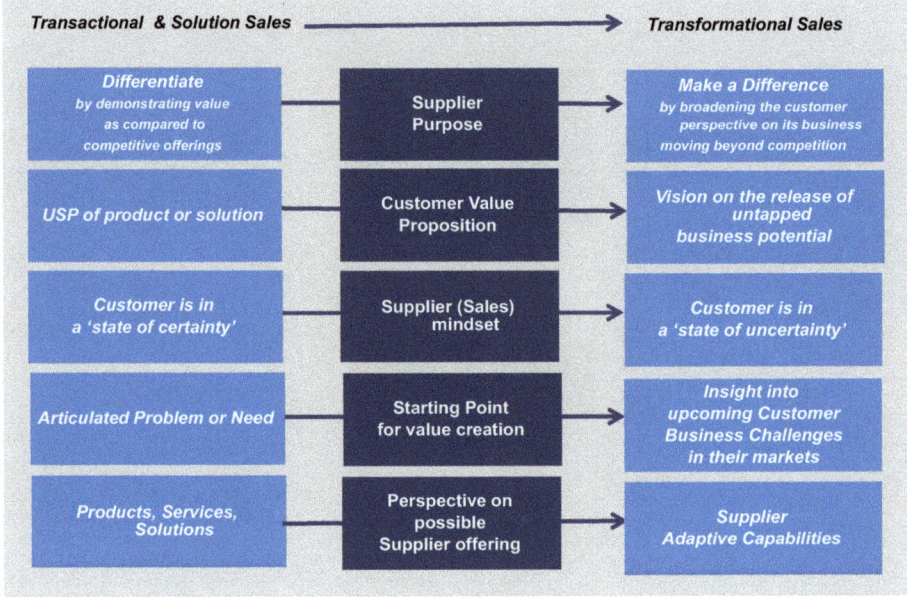

Fig. 2.12 Changing the customer's and supplier's thinking and way of doing business

the customer better than they understand themselves by taking new insights into the customer business challenges as a starting point for the conversation. As argued by Dixon and Adamson (2011, p. 52–53), the 'power of insight' addresses the customer need to 'learn' something rather than to 'buy' something.

This requires not only an 'outside-in understanding' of the changes in the customer world and the impact upon their business challenges. It also requires to redefine the buying center, identifying stakeholders 'open to change' in an earlier stage than the ones with the 'authority to spend' (Adamson et al. 2013). To be equipped to facilitate the release of the identified untapped potential within the customer business, the sales team needs an *intrapreneurial* perspective upon the available and accessible capabilities required and must think beyond existing products and services, beyond lines of business and regions and beyond traditional company borders. Furthermore it needs to define and acquire the 'adaptive capabilities' (Day 2011) required to create differential value in a changing world. In this sense transformational sales requires access to the supplier's total operation (Piercy 2010) and should be regarded as a business venture rather than as a sales initiative (Sherman et al. 2003). In essence we would like to define transformational sales as follows.

> **Transformational sales** is about guiding organizational change and business transformation, starting with sales itself. Companies choose this approach move beyond *product selling and problem solving* towards offering and provoking new *perspectives* upon the way of doing business. In essence transformational sales will transform both the customer and supplier businesses.

In the following chapters we will explore how to transform sales in order to transform the customer and supplier business.

References

Adamson, B., Dixon, M., & Toman, N. (2012). The End of Solution Sales. *Harvard Business Review, 90*(7/8), 61–68.

Adamson, B., Dixon, M., & Toman, N. (2013). Dismantling the Sales Machine. *Harvard Business Review, 91*(11), 103–109.

Ateş, M. A. (2014). *Purchasing and Supply Management at the Purchase Category Level: strategy, structure and performance*. Rotterdam: Erasmus Research Institute of Management (ERIM).

Axelsson, B., Rozemeijer, F., & Wynstra, F. (2005). *Developing Sourcing Capabilities*. Chichester: John Wiley & Sons, Ltd.

Bombardier. (2013). Financial Report Fiscal Year 2013.

Burnham, T. A., Frels, J. K., & Mahajan, V. (2003). Consumer Switching Costs: A Typology Antecedents and Consequences. *Academy of Marketing Science, 31*(2), 109–126.

Caniëls, M., & Gelderman, C. (2007). Power and interdependence in buyer supplier relationships: a purchasing portfolio approach. *Industrial Marketing Management, 36*(2), 219–229.

Cordón, C., & Vollmann, T. E. (2008). *The Power of Two. How Smart Companies Create Win-Win Customer-Supplier Partnerships That Outperform the Competition*. Basingstoke: Palgrave MacMillan.

Day, G. S. (2011). Closing the Marketing Capabilities Gap. *Journal of Marketing, 74*(4), 183–195.

Dingena, M. (2002). *Key Account Management*. Deventer: Kluwer.

Dixon, M., & Adamson, B. (2011). *The Challenger Sale. Taking Control of the Customer Conversation*. New York: Portfolio/Penguin.

Gosselin, D. P., & Heene, A. (2003). A competence-based analysis of account management; implications for a customer-focused organisation. *Journal of selling and major account management, 5*(1), 11–31.

Gupta, S., & Lehman, D. R. (2005). *Managing Customers as Investments*. Upper Saddle River: Wharton School Publishing.

Hardt, C. W., Reinecke, N., & Spiller, P. (2007). Inventing the 21st-Century purchasing organization. *The McKinsey Quarterly 4/2007* (October), 1–9.

Ingram, T. N., LaForge, R. W., & Leigh, T. W. (2002). Selling in the new millennium: a joint agenda. *Industrial Marketing Management, 31*, 559–567.

Kotler, P., Kartajaya, H., & Setiawan, I. (2010). *Marketing 3.0*. Hoboken: John Wiley &Sons.

Kraljic, P. (1983). Purchasing Must Become Supply Management. *Harvard Business Review, 61*(September/October), 109–117.

Kumar, V. (2007). How Valuable is Word of Mouth. *Harvard Business Review, 85*(10), 139–146.

Lindgreen, A., & Wynstra, F. (2005). Value in business markets: What do we know? Where are we going? *Industrial Marketing Management, 34*, 732–748.

Nichols, B., Sanders, B., & Mann, M. S. (2011). *The Journey to sales transformation: 25 Axioms for becoming a trusted partner to your customers*. San Bernhardino: Axiom Sales Force Development.

Peppers, D., & Rogers, M. (2011). *Managing Customer Relationships: A Strategic Framework*. New York: Wiley.

Peppers, D., & Rogers, M. (2012). *Extreme Trust, Honesty as a competitive advantage*. New York: Penguin.

Piercy, N. F. (2010). Evolution of Strategic Sales Organizations in Business-to-Business Marketing. *The Journal of Business & Industrial Marketing, 25*(5), 349–359.

Piercy, N. F., & Lane, N. (2009). *'Strategic Customer Management'*. Oxford: Oxford University Press.

Pine, B. J., & Gilmore, J. H. (2014). A Leader's Guide to Innovation in the Experience Economy. *Strategy & Leadership, 42*(1), 24–29.

Prahalad, C. K., & Krishnan, M. S. (2008). *The New Age of Innovation: Driving Co-Created Value through Global Networks*. New York: McGraw Hill.

Rackham, N., & De Vincentis, J. R. (1999). *Rethinking the Sales Force. Redefining Selling to Create and Capture Customer Value*. New York: McGraw-Hill.

Ritter, T., & Gemünden, H. G. (2003). Network Competence: Its Impact on Innovation Success and its Antecedents. *Journal of Business Research, 56*(9), 745–755.

Rozemeijer, F., Quintens, L., Wetzels, M., & Gelderman, C. (2012). Vision 20/20: Preparing Today for Tomorrow's Challenges. *Journal of Purchasing & Supply Management, 18*(2), 63–67.

Senn, C. (2012). The Booster Zone: How to Accelerate Growth with Strategic Customers. *Journal of Business Strategy*, *33*(6), 31–39.

Senn, C., Thoma, A., & Yip, G. S. (2013). Customer-Centric Leadership: How to Manage Strategic Customers as Assets in B2B Markets. *California Management Review*, *55*(3), 27–59.

Sherman, S., Sperry, J., & Reese, S. (2003). *The Seven Keys to Managing Strategic Accounts*. New York: McGraw-Hill.

Taleb, N. N. (2007). *The Black Swan*. London: Penguin Books.

Tarasi, C. O., Bolton, R. N., Hutt, M. D., & Walker, B. A. (2011). Balancing Risk and Return in a Customer Portfolio. *Journal of Marketing*, *75*(3), 1–17.

Terho, H. (2009). A Measure for Companies' Customer Portfolio Management. *Journal of Business-to-Business Marketing*, *16*(4), 374–411.

Verbeke, W., & Nagy, J. (2000). *Adaptief en strategisch accountmanagement*. Alphen aan de Rijn: Samsom.

van Weele, A. J. (2014). *Purchasing and Supply Chain Management. Analysis, Strategy, Planning and Practice*. Andover: Cengage Learning.

Wynstra, F. 2009, data from International Purchasing Survey Project (ipsurvey.org), provided by Rotterdam school of Management.

Wynstra, F. 2013, Lecture Rotterdam School of Management, Strategic Account Management Program.

Yip, G. S., & Bink, A. J. M. (2007a). Managing Global Accounts. *Harvard Business Review*, *85*(9), 103–111.

3
Setting the Joint Transformation Agenda

Instead of thinking outside-the-box. Get rid of the box

Deepak Chopra

Taking sales to a strategic level means moving away from the actual selling of existing products and services to engaging in business conversations with customers and other parties in the value network. It means being able to jointly strategize and discover value innovation opportunities and jointly setting the transformation agenda. This first of all requires a deep understanding of how the world around our customers is changing and influencing their real business challenges today and in the near future (see Sect. 3.1). To be able to connect to these customer business challenges, strategic sales teams also need to understand their own business in great depth (Piercy 2010; Sherman et al. 2003). In addition to **Customer Insight**, strategic sales teams need **Company Insight**. This means to truly understand the company's resources and competencies beyond current products and services sold. This requires the ability to think beyond the lines of business or product categories and to '[get] rid of the box' to see which resources are actually available or accessible either within the company or at value chain partners and might bring value to the table. This involves moving beyond solutions, inspiring customers with new insights into their business and making them more competitive in their end user markets (see Sect. 3.2). Value innovation opportunities can be discovered by bringing both worlds together, defining joint strategic focus with our strategic customers and jointly setting the transformation agenda to make the difference in the marketplace and move beyond the competition (see Sect. 3.3).

3.1 Customer Insight: Customer Business Relevance

> It is our goal to be one step ahead of our customers' demand (Martijn Legemaat, Corporate Account Insight Director, TNT).

One of the major strategic sales objectives is to discover new and longer term opportunities to create value both for the customer and the own organization. Starting to gain true Customer Insight requires starting to recognize future customer needs before they are made explicit by customers. This means moving beyond defined needs and problems as expressed by the customer towards the recognition that the customer is in 'a state of uncertainty' and converting the nature of the conversation towards 'disrupting the customer's thinking and assumptions about its business' (Adamson et al. 2013, p. 104). From a transformational perspective, Customer Insight means vocalizing potential customer business challenges and headaches based upon our observation and understanding of changes in the customer world; revealing major uncertainties that may impact customer business (e. g. Taleb 2007) and providing customers with new insights about their business and untapped business potential.

> **Transformational Customer Insight:** broadening the customer perspective on their business and their untapped business potential, by vocalizing the most relevant upcoming customer business challenges based upon our observation and understanding of changes in the customer world and their impact upon the customer business.

Transformational Customer Insight starts from the 'outside-in'; discovering how the world around our customers is changing by watching and understanding disruptive market trends. And more importantly by analyzing how these changes in the customer world will influence the customer's business challenges and headaches today and in the near future (see Fig. 3.1).

Making Sense of Developments in the Customer World

Developments in the customer world may move in a particular direction over time. Developments of longer duration are referred to as trends. Once developments are of shorter duration we may speak of 'hypes'. Trends may have a broad impact, influencing many aspects of society (macro or megatrends) or may be industry or customer specific (meso and micro trends). In many

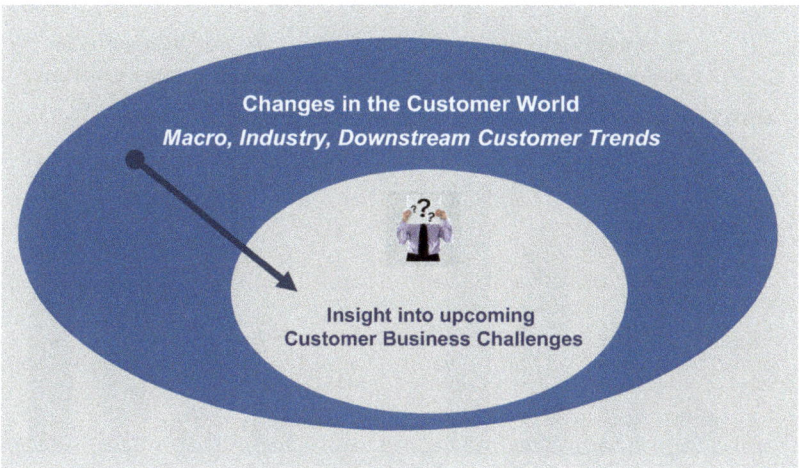

Fig. 3.1 Insight into upcoming customer business challenges

cases we see trends and anti-trends occurring within the same timeframe. For example, the trend of increasing (mobile) connectivity makes it possible to be (real-time) connected anytime and anywhere, enabling people to be '*always on*' (Boland et al. 2013). At the same time, we observe a growing interest in consciously slowing down (for example Slow Food and Slow Living), carefully selecting how to focus attention (e. g. Goleman 2013) and filtering what is relevant and what is not, consciously choosing a '*time out*' (Boland et al. 2013). The 'Slow Web Manifesto' for example calls to develop applications and technologies that guide us in filtering information, helping us to focus upon what is relevant and important again.

To be able to analyze changes in the customer world it is important to be aware of what and how we see. It is important to realize that our 'mental models' including assumptions and generalizations influence the way we try to make sense of market developments (Biemans 2010). Senge (1990) defines a 'mental model' as 'a series of deeply ingrained assumptions, generalizations, pictures and images that influence how an individual understands the world and how they take action' (As cited in Biemans 2010, p. 84). Our mental models do not only *color what* we look at (consider to be relevant), but also *how* we interpret market observations. Looking from the 'inside-out' may cause a 'myopic' view on the market (Levitt 1975; see also Dingena and Van Dishoeck 2002). Looking from the 'outside-in' may help us have a broad enough perspective to discover relevant and disruptive trends in the customer world. These trends may be analyzed on three levels: macro trends, meso or industry trends and micro trends.

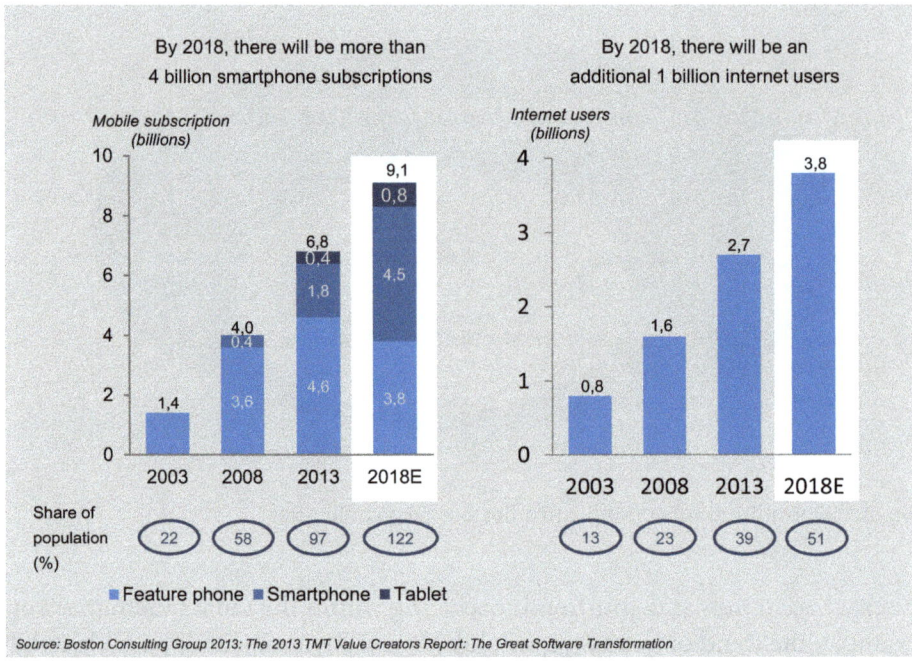

Fig. 3.2 BCG's value creators report: The global population is increasingly connected (Source: Boston Consulting Group 2013; reproduced with permission)

Macro Trends

On a macro level, an analysis can be made of the major **D**emographic, **E**conomic, **S**ociocultural, **T**echnological, **E**nvironmental and **P**olitical changes (DESTEP analysis). Which changes are going on and which of them will potentially disrupt our customers world in the coming years? For example, technological breakthroughs like mobile Internet, 3-D printing and Big Data analytics will impact many business challenges in the coming years. One of the major trends influencing all industries is the increasing (global) connectivity. As shown in Fig. 3.2, the global population is increasingly connected. It is predicted that by 2018 there will be more than 4 billion Smartphone subscriptions and around half of the global population will use the Internet.

Smartphones, smart devices and smart objects will be used to access remote information and connect with other devices, objects and processes, connecting the physical and digital world. This increasing connectivity will impact all industries, see for example the expected impact in travel.

Summarizing recent (online) trend reports and trend blogs (e. g. Boland et al. 2013, 2014; Boston Consulting Group 2013, Roland Berger 2014; Trendone 2014), the following macro trends can be identified: increased con-

> **Connected Travel**
>
> Travel will be different in many ways as a result of increased connectivity. Examples are:
>
> - The next generation of cars will be able to connect with smart devices, other cars, and with infrastructure.
> - Real-time travel information will be collected through voice- or gesture-controlled data glasses (or even contact lenses).
> - Big-Data analytics will enable quick changes across all modes of transport by processing real-time data.
> - By 2025 about 90 percent of all new cars sold worldwide will be able to communicate with infrastructure.
>
> Source: Industry reports, BCG-analysis as cited in Columbus 2014

Exhibit 3.1 Connected travel (Columbus 2014)

nectivity, increased use of smart devices and smart objects, growing importance of Big-Data analytics, merger of the physical and digital world (also referred to as phygital world), Industry 4.0 (see below), growing importance and application of 3-D (printing) technology, Human 2.0 (enabled by technological, genetic and spiritual developments), active aging, 'Age of access' (moving from 'owning' to 'using'; see also Sect. 4.1), endless ease (for example apps for everything and natural and playful user interfaces), increased inequality (see also the 'prosperity paradox' in the Shell scenario's in Exhibit 3.9; Piketty 2014), and an increased focus on the sustainable society and reconomy.

Industry 4.0

One of the fundamental changes taking place across many industries is the merger of the real (physical) world and the virtual (digital) reality. We are entering the so called 'phygital world' in many areas. This is visible in the world of production where physical objects and industrial processes are integrated into the information network, changing the rules of the game for industry players (Roland Berger 2014). This trend is referred to as 'Industry 4.0'.

Industry 4.0: The Fourth Industrial Revolution is Already on its Way

'Since the beginning of the 21st century, we have been experiencing a digital transformation-changes associated with innovation in the field of digital technology in all aspects of society and economy (…) This trend is also affecting the way goods are manufactured and services are offered. Western civilization has already witnessed three industrial revolutions, which could also be described as disruptive leaps in industrial processes resulting in significantly higher productivity. The first improved efficiency through the use of hydropower, the increasing use of steam power and the development of machine tools. The second brought electricity and mass production (assembly lines), and the third and most recent further accelerated automation using electronics and IT. The fourth industrial revolution is already on its way. This time, physical objects are seamlessly integrated into the information network. The internet is combining with intelligent machines, systems production and processes to form a sophisticated network. The real world is turning into a huge information system. 'Industry 4.0' provides the relevant answers to the fourth industrial revolution (…) Key characteristics of the new industrial landscape are:

- *Cyber-physical systems and marketplace*: In Industry 4.0 [IT] systems will be far more connected to all sub-systems, processes, internal and external objects, the supplier and customer networks (…) This enables highly efficient manufacturing in which production process can be changed at a short notice and downtime (e. g. at suppliers) can be offset.
- *Smart robots and machines*: Robots already replaced human workers in the last revolution (…) In industry 4.0, robots and humans will work hand in hand, on interlinking tasks and using smart sensored human-machine interfaces. The use of robots is widening to include various functions: production, logistics, office management (to distribute documents). These can be controlled remotely. If a problem occurs, the worker will receive a message on his mobile phone, with link to a web cam, so he can see the problems and give instructions to let the production continue until he comes back the next day. Thus the plant is operating 24 hours/day while workers are only there during the day.
- *Big Data:* data is often referred to as the raw material of the 21st century (…) A plant of the future will be producing a huge amount of data that needs to be saved, processed and analyzed (…) Innovative methods to handle big data and to tap the potential of cloud computing will create new ways to leverage information.
- *New quality of connectivity:* (…) In Internet 4.0 the digital and real worlds are connected. Machines, workpieces, systems and human beings will constantly exchange digital information via Internet protocol. This means physical things will be linked to their data footprint. Production with interconnected machines becomes very smooth: one machine is immediately informed when the part is produced in another machine, as well as the conveyor or the logistic supply robot (..). Even the product may communicate when it is produced – via an internet of things – and ask for a conveyor to be picked up. Or send an e-mail to the ordering system to say 'I am finished and ready to be delivered'. Plants are also interconnected in order to smoothly adjust production schedules among them and optimize capacity in a much better way.

- *Energy efficiency and decentralization:* Climate change and scarcity of resources are megatrends that will affect all Industry 4.0 players. These megatrends leverage energy decentralization for plants, triggering the need for the use of carbon-neutral technologies in manufacturing. Using renewable energies will be more financially attractive for companies. In the future, there may be many production sites that generate their own power, which will in turn have implications for infrastructure providers. In addition to renewable energy, decentralized nuclear power – e. g. small-size plants – is being studied as a way to supply big electro-intensive plants, thus providing double-digit energy savings.
- *Virtual industrialization:* (…) In Industry 4.0, we will use virtual plants and products to prepare the physical production; only once the final solution is ready the physical mapping done – meaning all software, parameters, numerical matrices are uploaded into the physical machines controlling the production. Some initial trials have made it possible to set up an automotive part production unit in three days – as opposed to the three months it requires today. Virtual plants can be designed and easily visualized in 3D as well as how the workers and machines will interact.'

Source: Roland Berger (2014, p. 7–9)

Exhibit 3.2 Industry 4.0: the fourth industrial revolution is already on its way (Roland Berger 2014, p. 7–9; reproduced with permission)

Meso or Industry Trends

On a meso level, the changes within the customer's industry, changes in 'industry logics' (Payne et al. 2008) and changing competitive forces may be analyzed. For example the '5 forces' model (Porter 1980) could be adopted and used to analyze changes within the customer industry (see Fig. 3.3).

In many cases, transformation within industries is driven by macro-trends. For example the connectivity trend tremendously influences the interaction between customers and suppliers in many industries. Also traditional industry borders are 'blurred' through new connections established as a result of increased connectivity. This brings in new entrants from previously unrelated industries, disrupting long-standing industry value chains. Both Google and Apple are exploring and developing 'self-driving car' possibilities that may reshape the automotive industry, travel and the way of living and working. Google officially announced the prototype of a driverless car in May 2014, whereas Apple launched their iCar platform in March 2014 (see Exhibit 3.3). At the same time, the power in the financial services industry may shift from traditional financial institutions towards innovative companies orchestrating alternative sources of (crowd) funding or providing the best digital (mobile)

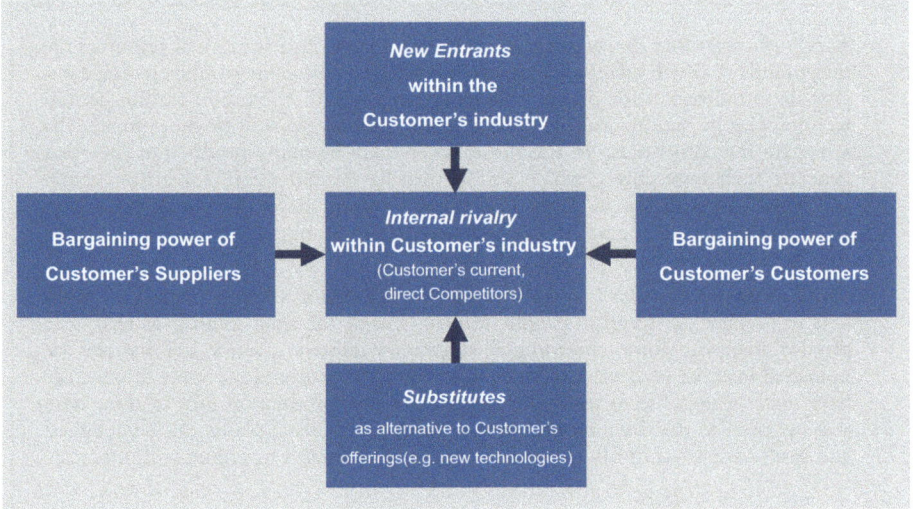

Fig. 3.3 Porter's 5 Forces driving competition within the customer's industry

wallets. The largest providers of business loans are not the traditional financial institutions, but instead companies like IBM and GE, offering financial solutions to their customers. Even not so large companies like the Trumpf Group, a € 3 billion metal forming and laser cutting machinery producer, has founded a general bank for the mechanical engineering sector to expand its service offering for the industrial clients.

As mentioned by Payne et al. (2008, p. 88) 'The blurring of industry borders and convergence of different types of industry represent opportunities to combine competences, capabilities and knowledge and initiate new ways of co-creating value'. This will not only lead to new entrants within the traditional customer industry, it will also open up opportunities for entering new areas of business together with our strategic customers.

Micro Trends: Changes in Downstream Customers Demand

On a micro level the changes in customer's customers' demand may be analyzed (e.g. Dingena 2010; Hillebrand and Biemans 2011; Anderson and Wouters 2013). Moving down the value chain and reasoning back may give useful insights into upcoming customer challenges, stemming from changes in downstream customers' demands. A company selling food ingredients may analyze changes at food retailers or end-consumers to initiate discussions with food producers about upcoming changes within their markets, thereby re-

> **Blurring Industry Borders: Apple Launches iCar Platform**
>
> 'Speculation over the launch of an Apple iCar has excited the tech industry for some time, and now it looks as if the wait is finally over (…) The tech giant unveiled a new platform that will enable drivers to integrate their iPhone and car entertainment system. The new product, called CarPlay, will be rolled out by Ferrari, Mercedes-Benz and Volvo. 'iPhone users always want their content at their fingertips and CarPlay lets drivers use their iPhone in the car with minimized distraction', said Greg Joswiak, Apple's vice president of iPhone and iOS Product Marketing. Through use of the new platform, drivers will be able to make calls, use the Google Maps function, listen to music and access messages through use of voice or touch. Users can control the program from the car's native interface or push and hold the voice control button on the steering wheel to activate Siri, the voice activation software. According to Apple, once a driver's iPhone is connected to a vehicle with CarPlay integration, Siri will enable drivers to access their contacts, make calls, return missed calls or listen to voicemails without using their hands. Drivers will be able to dictate responses to messages, or simply make a call. Apple said the product will also anticipate a driver's destination based on recent trips via contacts, emails or text and will provide routing instructions, traffic conditions and an estimated time of arrival. BMW Group, Ford, General Motors, Honda, Hyundai Motor Company, Jaguar Land Rover, Kia Motors, Mitsubishi Motors, Nissan Motor Company, PSA Peugeot Citroën, Subaru, Suzuki and Toyota Motor Corp are also set to offer CarPlay in the future. CarPlay will be available in selected cars shipping in 2014.'
>
> Source: CNBC news, 3 March 2014

Exhibit 3.3 Blurring industry borders: Apple launches iCar platform (CNBC News 2014)

defining food concepts. A chemical company may analyze developments in transportation to start conversations with converters and tire producers about joint innovations leading to safer driving. Anderson and Wouters (2013) mention, among others, suppliers of motion sensor technology used in full-body suits who incorporate information from customer's customers to improve their products and develop new applications. Analyzing dynamics at the customer's customers' level may not only result in adding new insights for strategic customers, it may also result in new forms of collaboration within or between value chains (see also Sect. 4.2). As argued by Anderson and Wouters (2013, p. 78) 'When three businesses in a value chain – a customer, its supplier and the customer's customer – work together to improve an innovation, the challenge of achieving a sufficiently shared understanding of technical detail to enable problem solving becomes greater than when just the customer and its supplier are working together'.

The importance to understand changes in customer's customers' demand is also confirmed by research applied from the purchasing perspective. According to Kibbeling et al. (2013), to realize superior value for end users it is of strategic relevance to develop relationships with suppliers that are 'able to anticipate environmental change through innovativeness' (Kibbeling et al. 2013, p. 512). As phrased by Mauricio Adade (Chief Commercial Officer of Royal DSM): 'we have the think B-to-C and act B-to-B' (see Exhibit 3.4).

Extended Decision Making Unit (DMU): Impact of Customer's Customers Demand

The value chain in which a strategic customer operates could be seen as an extended DMU. Strategic sales teams need to analyze influences beyond their immediate (direct) customer in the value chain to understand which parties beyond the customers influence decision making and in the end, who in the value chain contributes to the largest extent to decisions made (primary decision maker or primary customer within the value chain). See for example the extended analysis of the Decision Making Unit (DMU) within the value chain of a food ingredient manufacturer (see Fig. 3.4).

Initiating Conversations Challenging Customer Business Assumptions

Observed changes in the customer world may be used as a starting point for conversation with customers and can be jointly explored. Trends can be prioritized in terms of relevance and impact upon the customer business. In addition future scenarios or possible future worlds can be discussed based upon major uncertainties uncovered within the customer's market, thereby initiating conversations that enable the customer to explore business assumptions and imagine what future markets may look like by applying different 'lenses'. As mentioned by Shell (2013) it may be not very realistic to view changes in the customer world through one single lens, but instead develop different scenarios or lenses (see Exhibit 3.5).

Separating Customer Business Challenges and Headaches

In analyzing the impact of changes in the customer world upon the customer's business, a differentiation can be made in terms of customer 'business challenges' and 'business headaches' that stem from upcoming trends (Fig. 3.5). A *customer business challenge* could be defined as a future development or issue that 'excites' the customer about their future (where do they see opportunities

Royal DSM: Customer Insight Means 'Thinking B-to-C and Acting B-to-B'

Royal DSM is a global Life Sciences and Materials Sciences company active in health, nutrition and materials with 24,500 employees delivering an annual net sale of around € 10 billion. DSM delivers innovative solutions that nourish, protect and improve performance in global markets such as food and dietary supplements, personal care, feed, medical devices, automotive, paints, electrical & electronics, life protection, alternative energy and bio-based materials.

Mauricio Adade (Chief Marketing Officer of Royal DSM): 'Within DSM in the past years major developments took place in the area of strategic customer management. We revisited our key account management program and made a step forward to a customer business development approach. This included a selection of customers with whom we will be able to share strategies and jointly build strategic plans for the future. Jointly building and growing the business. This requires from our side to think beyond our direct customers. Crucial part of our customer business development approach is that we have to think 'B-to-C and act B-to-B'. We need to understand the complete value chain. Starting at the end of our value chain and reasoning backwards. Staying on top of new market trends, exploring unmet consumer needs and strategize around innovation opportunities that resonate with our business customers' marketeers. This joint approach resulted among others in an innovation award from our strategic customer Bayer (consumer healthcare division) in 2014. Our close collaboration resulted in increased customer attractiveness to Bayers' end consumers (as a consequence of innovative health ingredients) and reduced lead time in the supply chain. Additionally within our performance materials cluster this resulted in the successful collaboration between DSM Dyneema and strategic customers like Levi Strauss & Co (Durable Jeans) Reebok (Hockey Socks) and Ansell (breakthrough cut-resistant gloves). The cooperation with Levi's resulted in the launch of a stronger, longer lasting version of the classic Levi's 501®'.

DSM Dyneema and Levi's jointly create new market by renewing jeans user experience
The business group DSM Dyneema specializes in the manufacture of technologically sophisticated, high-quality products that are tailored to meet customers' performance criteria and used in a wide variety of end-use markets. Dyneema® is a super-strong fiber made from Ultra-High Molecular Weight Polyethylene (UHMwPE). The first applications, after DSM patented a process to spin the lightweight, high-polyethylene fiber from a gel process were ropes and cables, ballistic protection, and in composites. Over the past decades the combination of extreme strength and low weight appeared to be suitable for a wide and growing range of applications. It is used to protect people, anchor oil rigs, and harness wind power. When Levi's wanted to meet demands of people who wanted more durable, longer lasting jeans the joint project to create a new market for durable jeans started. As a result the Levi's 501® Strong jeans was launched. Jointly improving the consumer experience of target groups wearing their jeans longer.

Neil Bell (Global Fabric Innovation Manager at Levi Strauss & Co): 'Dyneema® is a man-made fiber designed for strength and durability. It's manufactured mainly for high-performance sports, industrial uses, and even for armor (...) Most of the other high-performance fabrics we've used in the past perform really well in their particular area. But with Dyneema®, we used only 4 percent and were able to get a very big impact and retain a cotton touch, which was critical. It's this cotton touch that we like – it feels like a pair of 501® jeans. We also don't need to use more cotton and materials to make more. We see the innovation of jeans with Dyneema® in the fact that they last two-

and-a-half times longer than a normal pair of jeans. By doing this, people don't need to buy a second pair of jeans (…) We did a partnership with DSM Dyneema because they claim to make the world's strongest fiber. We tested a lot of different fibers, but our strength testing proved that this is an amazingly strong fabric. It was a challenging project for us; we had to figure out how to spin this fiber in a completely different way. Normally what happens with these fibers is they are synthetic and shiny and it's very difficult to spin in a consistent way, so we had to take great care when we spun the yarn to make sure we really embedded this Dyneema® into the center of the fiber. That's really where the innovation was (…) Jeans look great and they feel comfortable. You can use them with any kind of apparel. You can use them in a casual or professional environment. And really, you shouldn't have to buy an extra pair or have designated pairs – one for casual and one for professional. I think jeans ought to cover both. And with Dyneema®, I think we've achieved that'.

Source: Interviews with Mauricio Adade (Chief Marketing Officer, DSM), Theo Verweerden (Marketing Program Director Value Creation, DSM), Rossana Rodriguez (Senior Marketing Consultant, DSM), November 2014; Company Presentation 2014, DSM at a Glance, DSM Factbook 2014. Interview with Neil Bell (Global Fabric Innovation Manager at Levi Strauss & Co) – published 8 January 2014.

Exhibit 3.4 Royal DSM: Customer Insight means 'thinking B-to-C and acting B-to-B' (based on interviews with Mauricio Adade (Chief Marketing Officer DSM), Theo Verweerden (Marketing Program Director Value Creation), Rossana Rodriguez (Senior Marketing Consultant, DSM) in November 2014; DSM 2014a, 2014b; Levi Strauss & Co 2014; published with permission)

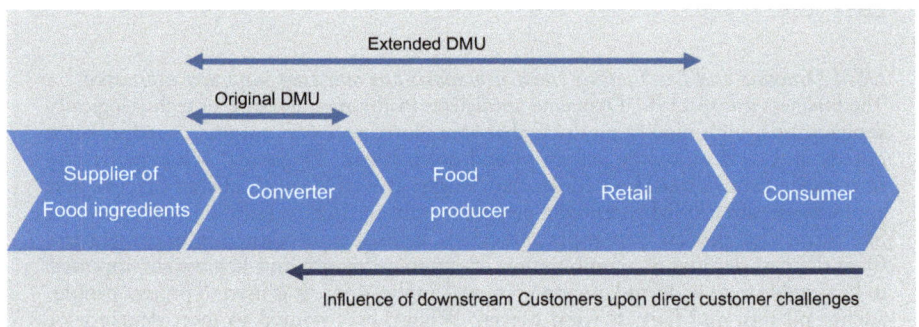

Fig. 3.4 Extended Decision Making Unit (DMU) of a supplier of food ingredients

and possibilities for growth or improvement?) A *business headache* could be defined as a future development or issue that the customer is worrying about, that keeps them awake at night (see also Cheverton 2008). It is vital to step entirely into the customer's shoes at this stage. The nature of the conversation will also give an indication to which extent customers are 'responsive to new insights about [their] business' (Adamson et al. 2013, p. 104).

> **New Lens Scenarios at Shell**
>
> *In an era of volatile transitions, it's unrealistic to propose a single lens through which to view the world of tomorrow.*
>
> *As our New Lens Scenarios note, by 2030 we expect demand for critical resources like water energy and food to have risen by 40–50 percent. To meet those needs without significant environmental detriment, business as usual will not be an option – we require business unusual.* (Peter Voser, CEO, Royal Dutch Shell)
>
> Shell is looking through the lenses of three paradoxes to highlight key features of the emerging landscape. These paradoxes are: the connectivity paradox, the prosperity paradox and the leadership paradox.
>
> *The connectivity paradox*
> Growing global connectivity stimulates creativity but also puts intellectual property at risk. Connectivity facilitates individual expression and empowerment, but also encourages herd behavior and amplifies swings in confidence and demand. The burgeoning availability of information has the capacity to bring insight and transparency, but data overload is equally likely to generate confusion and obscurity.
>
> *The prosperity paradox*
> Economic development is raising living standards for hundreds of millions of people. But it also imposes environmental, resource, financial, political, and social stresses that can undermine some of the benefits of prosperity. Private gains can flourish while public costs mount, and greater comforts today can lead to greater risks tomorrow. Globalization has tended to reduce income equalities between nations yet increases inequalities within them.
>
> *The leadership paradox*
> Addressing global stresses requires co-ordination among increasing constituencies of decision-makers. But the more diverse the groups that are involved, the more vested interests tend to block progress. An often-cited African proverb suggests that to go fast, go alone – but to go far, go together. Grappling with growing stresses requires that we go fast and far – implying a paradoxical need to go alone and together.
>
> Extract from the 'New Lens Scenarios' publication (Shell 2013)

Exhibit 3.5 New Lens Scenarios at Shell (Extract from Shell 2013; reproduced with permission)

Initiate Change with Selected 'Change Agents' in Preliminary Stages

We really work to try to drive to earlier, earlier, earlier development within our global account teams' (Fred Bell, Vice president of electronic manufacturing services at Molex Inc, as cited in Bartlett 2008).

Changes in the Customer World →	Impact on Customer Business	
	Business Challenges	Business Headaches
Macro trends		
Industry trends		
Downstream customer trends		

Fig. 3.5 Customer Insight: vision on untapped business potential

Pro-actively analyzing relevant upcoming business issues and challenging customer's assumptions about their business requires an entry to the customer decision making process before it is initiated by the customer (see Fig. 3.6). In other words moving towards transformational sales requires entering the conversation at a preliminary stage; taking the (pre)initiative to enter conversations with people open to change; vocalizing issues that are not yet articulated by the customer. Initiating these conversations also requires a redefinition of the customer's stakeholder selection. Instead of focusing on the decision makers and stakeholders with the 'authority to spend,' we need to focus upon stake-

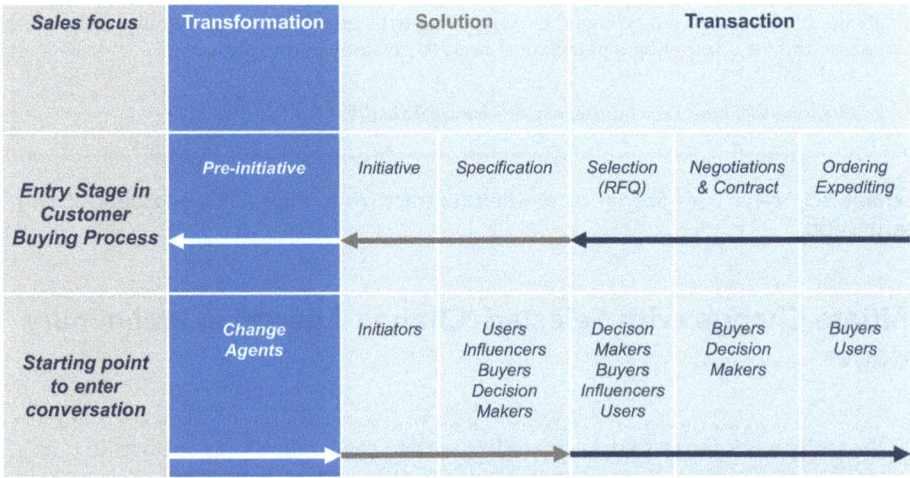

Fig. 3.6 Entering the conversation at a preliminary stage with selected change agents

holders who are 'open to change and can influence decision makers' (Adamson et al. 2013, p. 104). In many cases this requires broadening and redefining the customer buying center (DMU) to select the right 'change agents' within the customer organization (see Fig. 3.6).

3.2 Company Insight: Supplier Adaptive Capabilities

To be able to find a connection to customer business challenges the strategic sales team needs to understand their own business in great depth as well. In addition to Customer Insight, strategic sales teams need Company Insight. This means to have a profound insight in the own company's available and accessible resources and competencies beyond current products and services sold. These may be capabilities within the own organization and cross boundary, available within network partners and alliances. As argued by Blois and Ramirez (2006, p. 1030) 'The supplier must consider the potential contribution of any of the capabilities that it possess (...) Where a capability seems to possess the potential to make an appropriate contribution to the customer's value equation then there is a possibility that it can be used as a marketable asset'. Supplier capabilities may not only be seen as resources to build offerings for customers, but additionally as 'offerings or parts of offerings in their own right' (Blois and Ramirez 2006, p. 1027). Within the literature the term capability is used in different ways (see Blois and Ramirez (2006) for an overview.) We propose the following definition.

> **Supplier adaptive capabilities**: all available and accessible assets, resources and competencies that a company may use to facilitate or co-create customer value.

In Fig. 3.7 an overview of possible capabilities that could be used as marketable assets is shown.

As shown in Fig. 3.7, beyond existing products and services, capabilities can be available or acquired in the following areas.

Business and Market Intelligence Within the company and value network a large knowledge base is available that may be relevant in relation to the customer's business challenges. More and more the deployment of superior supplier knowledge and expertise is a 'defining characteristic of the world class sales organization, in the buyers' eyes. The buyer logic is straightforward: if the

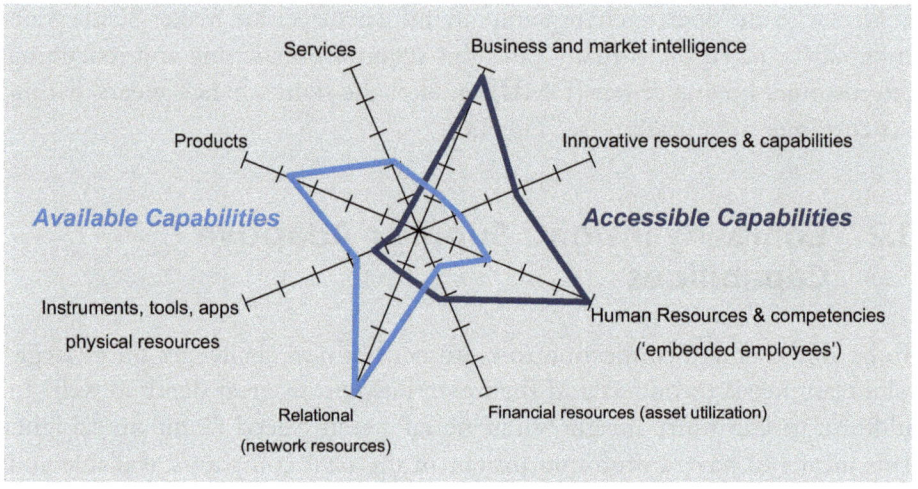

Fig. 3.7 Company Insight: Supplier Adaptive Capabilities

seller cannot bring value added to the relationship by identifying new opportunities for the buyer to gain competitive advantage in the end-use marketplace, then the seller is no more than a commodity supplier' (Piercy 2010, p. 353).

This superior knowledge and intelligence goes beyond information about products and solutions, which are sold today and require market intelligence about markets, regions and 'market sensing capabilities' (O'Cass and Ngo 2012). In addition business intelligence in other areas may be relevant to the customer such as strategic sourcing intelligence, technological information, financial and legal information and human resources information. Value can be created by disclosing this information and business intelligence and connecting it to customer business challenges, for example by using the knowledge and information in dialogues and by connecting customer counterparts with knowledgeable people within areas of the supplier organization or network that they have not been in contact with before.

Innovative resources and capabilities The capability to innovate is one of the essential capabilities to align to changes in the customer world and co-create value (O'Cass and Ngo 2012). These capabilities may include product, process and market innovation capabilities, by jointly developing new products and services, improving current processes and ways of working or finding innovative ways to enter or serve (new) customer markets as well as changing the rules in the marketplace. In addition available innovative resources may be used as an aid to guide transformation, such as testlabs, R&D results, studies, new (market) concepts, proven technologies, methods or intellectual property that could be licensed to customers.

> **Festo: 'Embedded Engineers' at Marel**
>
> Festo is a leading world-wide supplier of automation technology and the performance leader in industrial training and education programs with an annual sale of € 2.2 billion (2012). At Festo, 16,700 employees in 176 countries pursue a common goal: the maximum productivity and competitiveness of their customers in factory and process automation. Festo develops pneumatic and electrical drive and control technology for factory and process automation – from individual catalogue products to ready-to-install systems.
>
> In order to increase productivity and explore new ways to create value for customers like Marel (a leading global provider of equipment and systems for the food processing industry), Festo works with 'embedded engineers'. An embedded engineer works as a colleague among colleagues at the engineering (and/or R&D) department of a customer organization. Folkert Hettinga (Industrial Sales Manager Food & Beverage, Agriculture) said: 'Within Festo a dedicated team of 8 people is responsible to support Marel. This team includes people from sales, engineering, technical support, material management, back-office and our management team' One of our team members works as an 'embedded engineer' at the Engineering department of Marel. This provides us first-hand information on customer challenges in daily practice and fosters customer understanding and relations. Our embedded engineers leverage our potential to make a contribution to the customer business performance. At Marel the embedded engineer strengthens our role in the 'early supplier involvement program'. This has resulted for example in a faster construction of sub-assemblies, standardization, improved supply chain efficiency and just in time delivery of machine components'.
>
> Source: Interview with Folkert Hettinga (Industrial Sales Manager Food & Beverage, Agriculture at Festo), April 2014 and Festo Highlights 2014

Exhibit 3.6 Festo: 'Embedded engineers' at Marel (based on interviews with Folkert Hettinga (Industrial Sales Manager Food & Beverage, Agriculture at Festo), April 2014, and Festo Highlights 2014; published with permission)

Human Resources and embedded employees Human Resources like company specialists in various areas (e. g. technical, R&D, marketing, business intelligence) can be made available to strategic customers. These dedicated specialists may be either working at the company site (as a consultant, advisor, or assistant) or at the customer site. Dell Computer has 30 employees on-site at Boeing (using 100,000 Dell PC's) working closely together with Boeing managers in planning their requirements and in configuring the Boeing network (Anderson and Narus 2004). In industrial and technological surroundings we increasingly find 'residential' or 'embedded' engineers residing and working at the customer site, as part of the customer team, strengthening the customer-supplier interaction and feeding the supplier company with 'first-hand' information and ideas about possible value innovation (as illustrated in the example of embedded engineers of Festo at Marel in Exhibit 3.6).

> **Europcar and Daimler: car2go – On-Demand Mobility**
>
> Europcar is the leader in car rental services in Europe and one of the Top-3 Global players. The company serves business and leisure customers and creates flexible driving solutions which fit the different mobility needs of their target groups. According to Roland Keppler (CEO Europcar), the Europcar transformation agenda includes moving from a car rental company to a leading mobility player. Europcar's Fourth Annual Transportation & Mobility Observatory survey (under 6000 participants) shows that 'urban mobility needs evolve'. In fact 43 percent of European drivers, and even 47 percent of drivers in dense urban areas has considered giving up owning a car. This trend is particularly strong among drivers under the age of 35 and in higher socioeconomic categories. In response to these market changes, Europcar provides new services enabling customers to dispose freely of a car without owning one.
>
> *Payment per hour or per minute for on-demand car access*
> Together with DaimlerAG the car2go was launched in six cities in Europe in 2011 and early 2012. The car2go gives urban drivers the freedom and flexibility to rent a Smart car with no reservation or fixed rental period virtually anywhere within a metropolitan area. All that is required is to choose the available vehicle that is parked closest or most convenient. Payment can be by the hour or even the minute and the car can be dropped off simply by parking it.
>
> Esther van Koot (Commercial Director Europcar Netherlands): 'Within The Netherlands, in addition to car2go in the course of this year, a new mobility concept ('flexcar') will be introduced. Allowing customers to get access to a car whenever, wherever and for as long or as short as they want'.
>
> Source: Interview with Esther van Koot (Commercial Director Europcar Netherlands) May 2014 and Europcar Activity Report 2011–2012

Exhibit 3.7 Europcar and Daimler: car2go – on-demand mobility (based on interview with Esther van Koot (Commercial Director Europcar Netherlands) in May 2014 and Europcar Activity Report 2011–2012; published with permission)

Financial resources (asset utilization) Financial resources can be used in many ways to finance initial investments, provide loans, rent or lease or offer other financial or payment services. Financial resources may be used to optimize the customer's asset utilization. Also a growing number of on-demand services requires innovative payment concepts. In the last years we have seen a growing trend (referred to as the 'Age of Access' by Rifkin 2000) moving from owning to using, and from asset-based solutions to access-based offerings (see also Sect. 4.1). Europcar and Daimler jointly launched the car2go, offering on-demand car usage as an alternative for owning a car in dense urban areas (see Exhibit 3.7).

Relational resources and network competencies The quality of the relations built (relational advantage) and the ability to build relations and networks are at the core of building transformative customer-supplier relationships. The company's relevant capabilities are not limited to the internally available resources but depend on the relations developed and the network to which the company belongs (Castaldo 2007). These relational skills and resources or 'relational capital' (Blonska et al. 2013) are referred to as 'the sum of actual and potential resources embedded within, available through and derived from networks of relationships between organizations' (Nahaiet and Goshal 1998, as cited in Blonska et al. 2013, p. 1297). Strategic sales teams may leverage the collaborative relations built within the strategic customer organization as well as relations built within the value network to impact change or get access to relevant capabilities.

Available instruments, tools, apps or physical resources These may be diagnostics instruments, benchmark tools, (resource) planning tools, (mobile) web applications, marketing tools, portals, expert systems and so on. Any instrument that may provide customers with a better insight into their current business performance and give insight into untapped potential may bring value to the table.

Connecting the Dots

This is where the true *intrapreneurial spirit* comes into play (see also Sect. 5.2): being able to see potential connections between customer business challenges and one's own assets, resources and competencies. This can be realized either by using existing resources and or by enhancing, developing or acquiring capabilities within accessible reach. As argued by Day (2011), companies have to move beyond static activities to 'adaptive capabilities' that fit better to market reality. This requires thinking beyond lines of business or product categories to see which resources are actually available either within the company or within the value network and might bring value to the table. This involves getting rid of company constraints and offering new possibilities or connections, for example by connecting available technologies applied in particular markets to specific customer challenges in other markets or by exploring, redefining and developing new capabilities to address customer business challenges (see Exhibit 3.8).

A solid analysis of competencies and resources that could be used to address identified customer challenges can provide a basis for redefining value and give input for jointly developing the transformation agenda. See the example of ABInBev and JF Hillebrand in Exhibit 3.9.

Based upon a thorough Company Insight it is possible to find new connections to previously unaddressed customer business challenges and headaches.

Philips: Applying Natural Daylight Simulation Technology in Promising Areas

Royal Philips is a diversified health and well-being company, focused on improving people's lives through meaningful innovations in the areas of Healthcare, Consumer Lifestyle and Lighting. Headquartered in the Netherlands, Philips posted sales of € 23.3 billion in 2013 and employs approximately 115,000 employees with sales and services in more than 100 countries. Based on ongoing research and development, Philips is innovative in improving the lives of people by applying, for example, newly developed lighting technology. One area of research is 'Human Centric Lighting' which is intended to increase a person's well-being, mood and health.

Matthew Cobham (Lighting Application Team Manager, Indoor Professional Lighting Solutions, Europe): 'Philips is involved in research looking into how light can create an ideal environment for people. For example, by stimulating the natural sleep and wake cycle of people in the right way, by increasing concentration and by improving safety and efficiency in the workplace, healing, educational and work environments can be improved. To be more specific, the human body needs both the right light and darkness for the sleep-wake cycle to be optimized. Knowing when, for how long, the quantity and spectral content of the light is what needs to be understood'.

Originally the light technology was applied to increase the quality and duration of sleep and to improve people's mood. In consumer markets, daylight mimicking lamps are used to improve mood combined with an alarm clock to enable more natural waking up. Based upon the positive results in consumer lighting, further application of the technology has been investigated. For example, tests have been undertaken within the continuous production industry where night shifts are common. One of the major headaches of customers in this industry is the fact that productivity and experienced wellbeing tend to go down during the night. Providing the right light at the right time can result in significant increases in productivity combined with workers' wellbeing being positively influenced.

Other examples include education and healthcare. Research indicates that the adjustment of lighting in classrooms can have a direct effect on a child's ability to concentrate and learn. Under the dynamic daylight conditions not only did children's performance improve, but they also read faster and made fewer mistakes. Within Intensive Care units the biological effect of dynamic lighting supports the natural day and night rhythm of the patients and can enhance sleep quality.

Sources: Interviews with Selin Kelleci-Van Balen (Senior Regional Product Marketing Manager at Philips Lighting) and Matthew Cobham (Lighting Application Team Manager, Indoor Professional Lighting Solutions Europe) June 2014; Philips Annual report 2013; Philips 2013 (Schoolvision); Philips 2014; Lighting Europe/AT Kearney 2013

Exhibit 3.8 Philips: applying natural daylight simulation technology in promising areas (based on interviews with Selin Kelleci-Van Balen (Senior Regional Product Marketing Manager at Philips Lighting), Matthew Cobham (Lighting Application Team Manager, Indoor Professional Lighting Solutions Europe), June 2014; Philips 2013a, 2013b, 2014; published with permission)

ABInBev and JF Hillebrand: Redefined Value in Global Beverage Logistics

JF Hillebrand is a global logistics provider specialized in the beverage market. Combining expert staff, specialist knowledge and market-leading technologies, JF Hillebrand offers complete logistic management solutions to a range of customers across the globe. Strategic customers include large multinationals such as ABInBev, but also a wealth of wineries and other beverage producers, importers, retailers, wholesalers and traders; in total over 15,000 customers globally and 34,000 users of local services worldwide.

Tough start
For many years JF Hillebrand built relationships with companies like Beck's in Bremen, Stella Artois in Leuven, Spaten in Munich, and ended up with the largest brewer in the world, ABInBev. With the integration into the distribution network of ABInBev the volumes had shrunk for JF Hillebrand and with the centralization of transport and logistics procurement to the Headquarters in Leuven, the business was reduced to a minimum.

Moving along
Pierre Bonel (Chief Commercial Officer, JF Hillebrand): 'During discussions with ABInBev a few years ago, we learned that globally and across ABInBev's business service centers there was no established process for booking containers with forwarders and carriers. Furthermore, due to a lack of global tracking and monitoring tools, the procurement team and business units lacked the ability to track shipments, which impacted procurement quality. ABInBev had started to look for solutions via their IT vendors; however they soon realized that a different type of collaboration with their major logistics supplier would lead to a far better result. Gert van den Bossche (at that time Global Category Manager Ocean Freight at ABInBev) asked the determinant question: 'Why don't we do this together, aligning our infrastructure and working procedures?"

Transformative insight: redefining customer-supplier roles
A joint brainstorming session between ABInBev and JF Hillebrand resulted in the transformative insight that both the type of collaboration and joint value created could be different when addressed outside the traditional vendor-buyer framework. Both companies started to redefine their possible roles. At JF Hillebrand, Sales was challenging IT and Operations departments to deliver necessary adjustments to systems and processes to cater for the required customization. Joining efforts with the global procurement professionals at ABInBev, this new form of collaboration resulted in the ABInBev 'Ocean Control Tower'. A solution with strategic impact, increasing the efficiency of the order and information process and outsourcing part of the non-core shipping activity. Not only did this lead to cost reduction through better freight management, it also provided valuable data for supply chain improvements and internal logistics alignment. In addition, ABInBev's customer service quality significantly improved by the created dashboard to monitor KPIs such as lead-times, transit-times, demurrage, OTIF (on-time-in-full) and landed costs in an effortless way, thereby increasing the competitiveness of ABInBev in their end user markets. The co-created innovative solution outperformed the competition from global consultant and IT centered shipping-portals.

> ABInBev has retained full flexibility in handling freight procurement. JF Hillebrand helps in monitoring the transport vendors real-time with vendor KPIs, just as they would do for their own vendors. This intuitive combination of a transportation company and transparent procurement strategy resulted in increased value for both parties. Gert van den Bossche: 'It is getting better and better. With the Ocean Control Tower we can really control all our shipments and automatically track all the flows, globally. Also, the receiving countries (now more and more intercompany) will get more online info, to arrange for import formalities'.
>
> *Future collaboration*
> Sander Ouwehand (Current Corporate Account Manager at JF Hillebrand) is now further developing the relationship with ABInBev which is built on trust, an aligned collaborative approach and the co-creation of services: 'They share their 'workbook' now, which is actually their 3 year rolling strategic plan. This makes it much easier to jointly create value and build the future together'.
>
> Source: Interviews with Pierre Bonel (Chief Commercial Officer JF Hillebrand) and Sander Ouwehand (Corporate Account manager JF Hillebrand) in the period December 2013 to April 2014

Exhibit 3.9 ABInBev and JF Hillebrand: redefined value in Global Beverage Logistics (based on interviews with Pierre Bonel (Chief commercial Officer) and Sander Ouwehand (Corporate Accountmanager), December 2013–April 2014; published with permission)

Company Insight: exploring all available and accessible capabilities (to be acquired, developed, adapted or created) within the own company or within the company's value network that may be used to make the difference in addressing the most relevant customer business challenges or headaches.

3.3 Joint Strategic Focus

Joint strategic focus can be realized when the customer and supplier insights are connected. This is both a creative and strategic process. The customer/supplier collaboration matrix (adapted from Dingena 2002) can be a useful instrument for going through this joint process (see Fig. 3.8). Cheverton (2008) uses a similar instrument referred to as 'shared future analysis'. The aim is to analyze potential combinations between customer business challenges and supplier competencies and resources to reveal joint opportunities for value innovation. Joint analyzing of the matrix and choosing the most interesting opportunities to co-create value will not only result in joint strategic focus but will also increase commitment for the choices made. In fact a

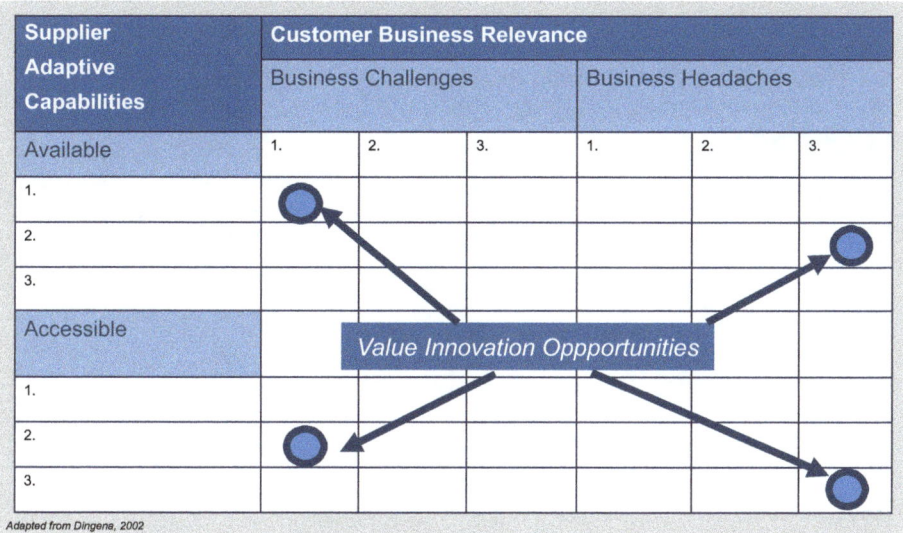

Fig. 3.8 Customer-supplier collaboration matrix: value innovation opportunities

'mental contract' is drawn up when joint choices are made and an agreement is reached on the most promising value innovation opportunities in the shared future.

Within the customer-supplier collaboration matrix prioritized customer business challenges and headaches are systematically related to the competencies and resources that could be used or developed from the supplier perspective. Doing this systematically will reveal the most interesting value innovation opportunities that may be put on the transformation agenda. Elaborating the matrix in a workshop setting with a multifunctional and cross boundary team (including people from both the customer and supplier organization) will enhance the objectivity of choices. Furthermore the joint elaboration increases commitment to the choices made (see also Dingena and Van Dishoeck (2002) for a detailed instruction about the use of the matrix).

The elaboration of the customer-supplier collaboration matrix may result in the selection of three to four strategic focus points (value innovation opportunities) on the joint transformation agenda (see Fig. 3.9 and also appendix B). Different value innovation opportunities emerge in the four different quadrants of the customer-supplier matrix resulting in four different areas of joint strategic focus (see Fig. 3.9): grasping a joint opportunity, joint development or innovation, joint defense or reduction of (perceived) risks and joint turnaround or radical business innovation.

Supplier Adaptive Capabilities	Customer Business Relevance	
	Customer Business Challenges	Customer Business Headaches
Supplier Available Capabilities	'Grasp joint opportunity'	'Joint defense/ reduction of (perceived) risk'
Supplier Accessible Capabilities	'Joint development/ innovation'	'Joint turnaround/ radical business transformation'

Adapted from Dingena, 2002

Fig. 3.9 Joint strategic focus

Grasping a Joint Opportunity

In this case a connection is selected between a customer business challenge and an existing relevant supplier capability to be used (more extensively). By joining forces the customer and supplier will be able to grasp a joint opportunity and grow together. Observed changes within customer markets or industries may result in growth opportunities in new market segments or regions where the supplier is currently active. By using or applying the supplier market knowledge, legal insights or the established network, the customer may be able to grow faster in this new market. Alternatively a customer may be faced with a changing demand in their markets, where the supplier may be able to provide the customer with insights or technologies building a first mover advantage within their market. This may result in jointly redefining standards within the industry.

Joint Defense or Reduction of (Perceived) Risk

In this case a connection is selected between a customer business headache and an existing relevant supplier competence or resource to be used (more extensively). By joining forces the supplier may support the customer to overcome or reduce the (perceived) problem or risk. Creating a new perspective on the worrying issue may be a starting point for change and applying existing technologies or knowing how that may have proven to be successful in other areas may also be used to ward off perceived headaches (see the before mentioned

> **Four Perspectives on Joint Innovation**
>
> - *Raise*: which elements, competencies or resources could be raised in the customer-supplier interaction to better address customer challenges?
> - *Create*: which new elements, competencies or resources could be developed (or acquired) in the customer-supplier interaction to better address customer challenges?
> - *Reduce*: which elements could be reduced in the customer-supplier interaction to better address customer challenges?
> - *Eliminate*: which elements could be eliminated in the customer-supplier interaction to better address customer challenges?
>
> Based upon: Kim and Mauborgne (1997, 2005)

Exhibit 3.10 Four perspectives on joint innovation (Kim and Mauborgne 1997, 2005)

Philips example applying the natural daylight simulation technology to reduce experienced business headaches within the continuous production industry).

Joint Development or Innovation

Another opportunity occurs when a customer business challenge is discovered and articulated, but the required resources or competencies are not (yet) available. Resources or competences may either be developed or acquired by cooperation with other parties. Joint development or innovation may make it possible to capitalize upon the discovered challenge. Siemens realized a joint development with its strategic customer Lufthansa to develop a wireless LAN for the airline (Senn 2006, p. 33). In many industries innovations are generated through the interaction between customers and suppliers. As commented on from the buying perspective: 'Increasingly, suppliers are an important source of innovation. The automotive industry may serve as an illustration. Innovations in fuel injection (Bosch), sun protection and security glass (Saint Gobain), retractable roofs (Inalfa), car seats (Lear), tire pressure sensors (Michelin) and airbags (Autoliv) came from suppliers' (Van Weele 2014, p. 215).

It is important to note that joint innovation may be realized both by increasing or creating new value elements in the supplier-customer interaction, but also by reducing or even eliminating value elements (e. g. Kim and Mauborgne 1997, 2005; see Exhibit 3.10). Reducing elements in the offering may result in 'naked solutions or naked systems' (Anderson and Narus 1995). 'ABB found that naked solutions enabled it to charge less for power equipment and heavy industrial equipment than it could for the standard package designed for the average customer' (Anderson and Narus 1995, p. 76). Eliminating elements or

> **GE's Quest Program: 3D Printing Quest to Improve Efficiencies in Healthcare Industries**
>
> GE mobilizes the intelligence of the crowd to disrupt current thinking. As Steve Liguori (Executive Director of Global Innovation at GE) said: 'Harnessing the power of the crowd is essential in disrupting current processes and accelerating the pace of innovation. GE's Quest program taps into the world's greatest minds to create products that bring new values to our customers and speeds the time from mind to market'. GE's Quest program challenged data scientists, academia, start-ups and established business worldwide to use analytics and advanced manufacturing processes to find ways to increase efficiency for airlines and healthcare customers.
>
> GE's 3D Printing Production Quest, in partnership with NineSigma, challenged participants to use additive manufacturing to produce complex parts with high precision using refractory metals. This capability could transform how components are manufactured for X-ray-based medical imaging systems such as mammography, cardiac catheterization and computed tomography. As the global medical imaging market is expected to reach $ 35.35 billion by 2019, GE envisions additive manufacturing enabling new component designs that greatly simplify manufacturing and reduce cost, while improving image quality and diagnostic capability.
>
> Refractory metals have a high density allowing them to very effectively block X-rays without the environmental and health hazards associated with lead, and also have very high melting temperatures. They are used in X-ray systems to control the path of X-rays from the source through the patient's body and some components such as X-ray source tubes that take advantage of the high melting temperature.
>
> Source: GE (22 April 2014)

Exhibit 3.11 GE's Quest Program: 3D Printing Quest to improve efficiencies in healthcare industries (GE 2014)

products, also referred to as 'creative destruction,' may additionally be a way to increase value in a changing market. The choice exists to eliminate what is there to make place for new alternatives which are a better fit to changing market circumstances. In many cases this means eliminating existing products and services before new entrants within the industry take over.

Joint Turnaround or Radical Business Transformation

The largest transformation in the customer-supplier interaction will occur in the situation where a connection is chosen between a customer headache and the search for new resources or competencies to be developed, changed or acquired. In many cases this requires not only a redefinition of the customer and supplier roles but includes a different way of working within the value network. This may include joining forces with competitors or other parties in the value network (see also Sect. 4.3). Increasingly, also crowd sourcing and collective intelligence are used to boost radical business transformation (see for example GE's Quest Program in Exhibit 3.11).

The joint transformation agenda defines the strategic focus in the customer-supplier relation. The various ways to actually impact the business results and guide business transformations are elaborated in more depth in Chap. 4.

References

Adamson, B., Dixon, M., & Toman, N. (2013). 'Dismantling the Sales Machine'. *Harvard Business Review, 91*(11), 103–109.

Anderson, J. C., & Narus, J. A. (1995). Capturing Value of Supplementary Services. *Harvard Business Review, 73*(1), 75–83.

Anderson, J. C., & Narus, J. A. (2004). *Business Market Management. Understanding, Creating and Delivering Value*. Upper Saddle River: Pearson, Prentice Hall.

Anderson, J. C., & Wouters, M. (2013). What You Can Learn From Your Customer's Customer. *MIT Sloan Management Review, 54*(2), 75–82.

Bartlett, G. (2008). The Global Account Manager at Molex Inc: decision maker, value creator and strategist. *Velocity – Strategic Account Management Association, 10*(Q1), 31–34.

Biemans, W. G. (2010). *Business to business marketing: a value-driven approach*. London: McGraw-Hill Higher Education.

Blois, K., & Ramirez, R. (2006). Capabilities as Marketable Assets: A Proposal for a Functional Categorization. *Industrial Marketing Management, 35*(8), 1027–1035.

Blonska, A., Storey, C., Rozemeijer, F., Wetzels, M., & de Ruyter, K. (2013). Decomposing the effect of supplier development on relationship benefits: the role of relational capital. *Industrial Marketing Management, 42*, 1295–1306.

Boland, C., Bosma, T., Bullinga, M., Eilander, G., Kniesmeijer, T., Lamb, R., Mirani, N., Rohde, C., Roothart, H., Schönfeldt, E., Van den Hoff, R., & Witmer, M. (2013). *Trendrede 2013*

Boland, C., Bosma, T., Bullinga, M., Eilander, G., Kniesmeijer, T., Lamb, R., Mirani, N., Rohde, C., Roothart, H., Schönfeldt, E., Van den Hoff, R., & Witmer, M. (2014). *Trendrede 2014*

Boston Consulting Group (2013). *The 2013 TMT Value Creators Report: The Great Software Transformation, How to Win as Technology Changes the World*. Boston: Boston Consulting Group.

Castaldo, S. (2007). *Trust in Market Relationships*. Cheltenham: Edward Elgar Publishing.

Cheverton, P. (2008). *Global Account Management*. London: Kogan Page.

CNBC News (2014). Has the iCar arrived? Apple launches in-car platform. *CNBC news*, Monday 3 March 2014.

Columbus, L. (2014). *BCG's Value Creator's Report Shows How Software is driving new business models, January 2014*. Softwarestrategiesblog.com

Day, G. S. (2011). Closing the Marketing Capabilities Gap. *Journal of Marketing*, *74*(4), 183–195.

Dingena, M. (2002). *Key Account Management*. Deventer: Kluwer.

Dingena, M. (2010). Focus on the Customer's Customer. *Kendrion Magnetized*, *02*(Q3), 7.

Dingena, M., & Van Dishoeck, N. (2002). *Successful Marketingplanning*. Boulder: F&G Publishing.

DSM, 2014a. *Company Presentation 2014. A short introduction to DSM*.

DSM, 2014b. *DSM At a Glance. Factbook 2014*.

Europcar Activity Report, 2011–2012.

Festo Highlights, 2014.

GE, 2014, 'GE and partners announce winning 'open collaboration' innovations of Industrial Internet Flight Quest 2 and 3D Printing Production Quest', April 22, 2014.

Goleman, D. (2013). The Focused Leader. *Harvard Business Review, 91*(12), 51–60.

Hillebrand, B., & Biemans, W. G. (2011). Dealing with Downstream Customers: An Exploratory Study. *Journal of Business &amp Industrial Marketing*, *26*(2), 72–80.

Kibbeling, M., v. d. Bij, H. H., & v. Weele, A. (2013). Market Orientation and Innovativeness in Supply Chains: Supplier's Impact on Customer Satisfaction. *Journal of Product Innovation Management*, *30*(3), 500–515.

Kim, C. W., & Mauborgne, R. (1997). Value Innovation: The Strategic Logic of High Growth. *Harvard Business Review, 75*(1), 103–112.

Kim, C. W., & Mauborgne, R. (2005). *Blue Ocean Strategy*. Boston: Harvard Business School Press.

Levi Strauss & Co, 2014. 'Sourcing Dyneema's super strength: an interview with LS&Co fabric innovator Neill Bell', 8 January 2014.

Levitt, Th. (1975). 'Marketing Myopia' (With retrospective commentary). *Harvard Business Review* (September/October), 26–48. (original 1960)

O'Cass, A., & Ngo, L. V. (2012). Creating Superior Customer Value for B2B Firms through Supplier Firm Capabilities. *Industrial Marketing Management*, *41*(1), 125–135.

Payne, A. F., Stobacka, K., & Frow, P. (2008). Managing the Co-Creation of Value. *Journal of Academic Marketing Science*, *36*(1), 83–96.

Philips, 2013a, Annual report 2013.

Philips, 2013b, School vision.

Philips, 2014 – Lighting Europe/At Kearney – 2013.

Piercy, N. F. (2010). Evolution of Strategic Sales Organizations in Business-to-Business Marketing. *The Journal of Business & Industrial Marketing*, *25*(5), 349–359.

Piketty, T. (2014). *Capital in the Twenty-First Century*. Cambridge: The Belknap Press of Harvard University Press.

Porter, M. E. (1980). *Competitive Strategy*. New York: Free Press.

Rifkin, J. (2000). *The Age of Access. The New Culture of Hypercapitalism where all life is a Paid for experience*. New York: Jeremy P. Tarcher, Putnam, a member of Penguin Putnam Inc.

Roland Berger (2014). *Industry 4.0. The new industrial revolution. How Europe will succeed* vol. March 2014.

Senge, P. (1990). *The Fifth Discipline: The Art and Practice of the Learning Organization*. New York: Doubleday

Senn, C. (2006). The Executive Growth Factor: How Siemens Invigorated its Customer Relationships. *The Journal of Business Strategy*, *27*(1), 27–34.

Shell (2013). *New Lens Scenarios. A shift in perspective for a world in transition.* : Shell International.

Sherman, S., Sperry, J., & Reese, S. (2003). *The Seven Keys to Managing Strategic Accounts*. New York: McGraw-Hill.

Taleb, N. N. (2007). *The Black Swan*. London: Penguin Books.

Trendone (2014). *Trendbook 2015. Das Zukunftslexikon der wichtigsten Trendbegriffe*. Hamburg: Trendne.

van Weele, A. J. (2014). *Purchasing and Supply Chain Management. Analysis, Strategy, Planning and Practice*. Andover: Cengage Learning.

4
Guiding Customer Business Transformation

What you do makes a difference, and
you have to decide what kind of difference you want to make
                                                          Jane Goodall

Guiding business transformation for the customer is about increasing customer value by guiding 'company-altering' or 'behavior-altering' experiences (Pine and Gilmore 2014) and increasing value in such a way that customers become more competitive and successful in their markets. Customer value has been defined in many ways. See for example Sánchez-Fernández and Iniesta-Bonillo (2007) for an extensive overview. They mention that the most commonly cited definition is based on Zeithaml (1988), who refers to value as the assessment of 'what is received and what is given', rephrased by Sánchez-Fernández and Iniesta-Bonillo (2007, p. 428) as a trade-off between 'benefit and sacrifice'. Even though the value concept in business markets has been further elaborated by many authors, we prefer to take this essential definition as a starting point: *'Value in business markets is a relationship between what one benefits and what one sacrifices'*. From this perspective, value to customers means that perceived benefits of doing business with a chosen supplier are larger than the perceived sacrifices or costs. In other words customers are and/or feel better off after the interaction with a supplier than before (Grönroos 2011). The customer's business 'well-being' is increased and the customer becomes 'better off in some respect' (Grönroos and Voima 2013). Obviously both perceived benefits and sacrifices or costs go beyond the economic functions which a particular offering may fulfill. In addition to time, money and function they may include emotional aspects like perceived risk, fear, frustration, comfort, peace of mind, inspiration, trust and so on.

Increasing customer competitiveness in their markets starts with understanding and relating to the customer's 'competitive priorities' (Ateş 2014). Customers may primarily be focusing on the reduction of cost levels or (directly or indirectly) on the increase of their revenue streams or both. The actual supplier impact can be either economic (financial or functional) or emotional. Customer business transformation (increasing customer competitiveness and success in their markets) can therefore be approached from four possible perspectives, i. e. decreased economic or emotional costs and/or or increased economic or emotional revenues.

> **Customer Business Transformation** means increasing customer competitiveness and success in their markets by decreased economic or emotional costs and/or increased economic or emotional revenues.

In this chapter, we will explore in depth these four perspectives on customer business transformation. Within these four perspectives, eight different ways to guide business-altering experiences are distinguished (see Sect. 4.1).

One way to significantly improve customers' competitive advantage is by building new connections, joining forces with partners, alliances, suppliers or even competitors to change the way in which value is created. In times of increasing connectivity we see a change from linear value chains to nonlinear value networks. Suppliers can act as a 'lead collaborator' (Vitale et al. 2011) in these integrated (global) value networks (see Sect. 4.2).

To visualize and vocalize the essence of what the increased competitiveness will look like, it is vital to craft and use 'resonating' value propositions (Anderson et al. 2006). These value propositions have to connect and touch upon what is most relevant to the customer; touching upon the untapped business potential within the customer organization and personalized to key stakeholders; clarifying how the supplier will contribute and make a difference. In addition, these propositions need to be vocalized in a way the customer feels and understands, building a rational and emotional connection. This includes speaking the customer (business) language; for example by elaborating 'value bridges' and calculating and communicating the impact to the customer in monetary terms (Anderson et al. 2008) and in the customer 'money-making logic' or related to other customer value drivers (see Sect. 4.3).

4.1 Making Customers More Successful in Their Markets

Transformational sales is about making customers more competitive and successful in their markets. As highlighted by Ateş (2014), customers may set different 'competitive priorities' within their operations and purchasing strategies to increase their competitive advantage in their markets. One priority may be to reduce cost levels or reduce perceived sacrifices. Another priority may be to increase revenues or strengthen competencies to increase revenues such as improved time-to-market or innovation. According to Grönroos (2011, p. 42) the value a 'customer can create out of the support provided by a supplier can be divided into three dimensions':

- Effect on the customer's growth- and revenue generating capacity. Grönroos divides these into business growth opportunities (e. g. new markets, better customer or segment penetration) and higher margins through premium pricing.
- Effect on customer's cost level by lowering operative or administrative costs or higher margins through lowering operating and/or administrative costs.
- Effect on perceptions such as trust, commitment and comfort in supplier interactions and increased attraction of the supplier.

Four Windows of Opportunity

In essence a customer's competitiveness may be increased by focusing on two dimensions. First a focus on the customer's competitive priorities (see Ateş 2014), which may be either to reduce cost levels or to – directly or indirectly – increase revenues (or both). Second a focus on the actual supplier impact in the customer organization which may be either economic (financial or functional) or emotional.

In addition to the economic and emotional impact, the environmental or *ecological* and socially responsible impact may be distinguished. Supporting customers to build a sustainable business and improve the ability to transform the world around them in a positive way. This may be one of the crucial themes to make a meaningful difference in the marketplace in the coming years and to motivate employees and business partners at the same time. See for example IBM who before the full rollout of grid computing to commercial customers gave the technology to scientists (see Exhibit 4.1).

Even though sustainable and socially responsible purchasing seem to become increasingly important on the purchasing agenda as well (see also Sect. 2.2), in practice we observe a close correlation with economic output. Companies focus on sustainable *and* profitable possibilities and business models rather than sustainable solutions per se. In the model in Fig. 4.1 we have chosen to include the ecological and social responsible impact as one of the parameters of economic impact, rather than to treat it as a separate variable.

In Fig. 4.1 we distinguish the economic and emotional impact. The economic impact can be demonstrated in monetary terms (e. g. Anderson et al. 2008). Emotional impact is about how the interaction (and the result of the interaction) makes the customer feel and about the closeness or the intimacy

> **IBM: Offering Business Partners an Opportunity to Contribute to Something Big**
>
> 'At IBM, even before the rollout of grid computing to commercial customers, the company gave away the technology to scientists searching for cues for HIV-AIDS, heart disease, and cancer. Grid computing enables the aggregation of individual PC's power through a network, providing the information processing necessary for big, ambitious research. As soon as IBM had perfected the technology, it created a nonprofit partnership, World Community Grid (WCG), through which any organization or individual can donate unused computing power to research projects and see what is being done with the donation in real time. Through WCG, IBM gains an inspiring showcase for its new technology, helps business partners connect with the company in a positive way, and gives individuals anywhere in the world the chance to contribute to something big'.
>
> Source: Moss Kanter (2008, p. 44)

Exhibit 4.1 IBM: offering business partners an opportunity to contribute to something big (Moss Kanter 2008, p. 44)

Customer Business Transformation	Customer Competitive priorities	
	Cost	Revenue
Economic	Decreased Economic Costs	Increased Economic Revenues
Emotional	Decreased Emotional Costs	Increased Emotional Revenues

(Supplier Impact)

Fig. 4.1 Making customers more successful in their markets: four windows of opportunity

of the relationship. Even though the emotional impact may be more difficult to quantify directly in monetary terms it is fundamental to the actual and possible (future) value innovation and impacts profitability in many ways (e. g. Castaldo 2007). In practice strategic sales teams may approach the customer business transformation from four perspectives or windows of opportunity: decreased economic costs, decreased emotional costs, increased economic revues and/or increased emotional revenues (see Fig. 4.1).

> **Dell: Streamlining Processes by Moving Unilever's IT Deployment into Dell's Factory**
>
> Dell observed how large customers frequently purchased hundreds of computers simultaneously, after which their IT departments spent days installing the required software and distributing them to the desired locations. During this process, the computers were not used, IT employees are full-time installing software and hallways were filled with empty boxes for days. Dell streamlined processes, resulting in a decrease of 30 percent of the customer's deployment work in Dell's factory. For Unilever, Dell streamlined the roll-out of 10,000 pre-imaged Dell OptiPlex desktops and Dell Latitude laptops with Intel processors running on Microsoft Windows across over 100 locations. As concluded by Roger Legendre, Unilever's Director of Technology Solutions Deployment: 'Dell managed the necessary vendors for us and made sure that we had pre-imaged OptiPlex desktops and Latitude notebooks on site and ready to be installed and tested within the narrow timeframe'.
>
> Source: Biemans (2010, p. 113)

Exhibit 4.2 Dell: streamlining processes by moving Unilever's IT deployment into Dell's factory (Biemans 2010, p. 113)

Decreased Economic Costs

A first area to increase competitiveness is to decrease monetary costs beyond lowering the purchasing price. This can be realized by reducing *cost levels* in primary or operative customer processes and in supporting or administrative processes, by focusing on the decrease of total cost of ownership or focusing upon the upon the optimization of asset utilization. Suppliers can for example enable the streamlining of processes by insourcing some of the activities previously carried bout in the customer organization (see for example Dell, streamlining processes of strategic customer Unlever in Exhibit 4.2).

Decreased Emotional Costs

In many cases it is additionally worthwhile to critically assess the possibilities to decrease emotional costs or sacrifices in customer-supplier interactions. As argued by Pine and Gilmore (2014, p. 24): 'Every business would benefit from asking itself: What one dimension of sacrifice, if eliminated, would create the greatest value for our customers?' Emotional costs or sacrifices perceived in business interactions with suppliers may include (perceived) risk levels and other psychological costs like frustration, disappointment, anger, irritation, nervousness or even anxiety.

> **Europcar Moving Your Way: Flawless Experience for Business Travelers**
>
> Europcar is the leader in car rental services in Europe and one of the Top-3 Global players. The company serves business and leisure customers and creates flexible driving solutions which fit the different mobility needs of their different target groups. Based upon Europcar's Promotor score survey, valuable customer input is gained in order to improve the customer experience. The survey is used among other things to identify elements that may prevent a customer from having a flawless experience. In other words to reduce or eliminate perceived 'sacrifices' by the customer. 'The systematic analysis of customer comments about their rental experience provides a wealth of information, enabling Europcar staff throughout the organization to initiate improvements'. For the target group of international business travelers this resulted in a streamlined reservation, pick-up and drop-off process. Reducing not only time spent in hiring, picking up and dropping off a car, but also reducing experienced (pre)purchase risk for example by offering a 'get to your destination guarantee' and 'trouble free service guarantee'. In addition, experienced inflexibility was decreased by offering variability in drop off locations and combined rental car and driver services. Creating a flawless experience for business travelers at international airports and elsewhere.
>
> Source: Interview with Esther van Koot (Commercial Director Europcar Netherlands) May 2014 and Europcar Activity Report 2011–2012

Exhibit 4.3 Europcar moving your way: flawless experience for business travelers (based on interview with Esther van Koot (Commercial Director Europcar Netherlands) and Europcar Activity Report 2011–2012; published with permission)

Increased Economic Revenues

A third and less explored area in customer-supplier relations is the potential to increase customer's competitiveness by the increase of economic revenues. Customer revenue growth may be directly increased by connecting customers to new (joint) business opportunities or by offering them reciprocal value opportunities. In addition, economic revenues may be indirectly increased by facilitating the customer in increasing their 'revenue generating capacity' (Grönroos 2011) and supporting the customer in addressing new market demands or with development and the introduction of new products or services. Other areas include increasing time-to-market, improving product quality, delivery reliability, flexibility, responsiveness to customer's customers or other ways to increase the customer's customers' satisfaction. Also attention for increased sustainability (ecological impact) may have a positive impact on the increase of economic revenues.

Making a Difference		Customer Competitive priorities	
		Cost	Revenue
Supplier Impact	Emotional	1. Decreased negative emotions	3. Increased positive emotions
		2. Decreased risk	4. Increased trust
	Economic	5. Decreased Total Cost of Ownership	7. Increased revenue generating capacity
		6. Optimized asset utilization	8. Increased revenues

Fig. 4.2 Business impact: eight ways to make a difference to the customer business

Increased Emotional Revenues

> I've learned that people will forget what you said, people will forget what you did, but people will never forget how you made them feel (Maya Angelou).

The fourth and most fundamental area to increase a customer's competitiveness is by increasing the emotional revenues within the customer-supplier relationship. Either by increasing trust within the customer organization or by increasing other emotional benefits such as increased comfort, peace of mind, interest, hope, inspiration, pride or happiness with the major stakeholders. We believe that the increase in emotional revenues or in other words, the improved emotional connection within the customer organization is prerequisite to gaining real progress in any of the other areas.

Eight Ways to Make a Difference Beyond Lowering Your Price

Within the four windows of opportunity we may distinguish at least eight different ways to enable business-altering experiences and guide customer business transformation (see Fig. 4.2). The emotional connection and built trust are seen as a threshold to profound economic impact, requiring open information sharing between the customer and supplier organizations.

1. *Decreased negative emotions such as frustration, fear, and anger*

An often underestimated area to increase value is the reduction of emotional sacrifices or costs a customer may perceive in the interaction with the supplier.

These emotional or psychological costs may be related to any negative emotions experienced pre-purchase and in the relation and interaction with the supplier. This may not seem a big issue when related to standard products or solutions. However, when innovative collaborative projects are at stake (e. g. the development of new food ingredients, new market concepts, water turbines or power plant cooling systems), these emotional factors are enormous and of significant importance. Customers may feel worried or insecure about the actual performance or timing of a supplier's offering or experience a lack of control during usage. Also insufficient management of expectations may lead to disappointment or frustration or even anger and fear. Innovative concepts may affect feelings of insecurity or anxiety with decision makers and users in the customer organization. Negative emotions can not only impact the customer-supplier relationship in a negative way, they can also result in additional unnecessary 'indirect relationship costs' like extra time spent double checking or lost business opportunities (Grönroos 2007). Therefore decreased emotional costs will not only improve the customer-supplier relationship; they will also increase customer competitiveness. It may be worthwhile to literally step into the customer's shoes and explore the customer experience in the total 'customer-activity-cycle' to understand where emotional costs are experienced or may be experienced in future collaboration.

2. Decreased (perceived) risk

To be able to decrease (perceived) customer risk, it is useful to analyze the sources of the perceived risk in the customer-supplier interaction. Perceived risk may stem from the actual availability of raw materials, goods and services, the number of suppliers, competitive demand, make-or buy opportunities, storage risks (in case of physical supplies) and substitution possibilities related to purchased and used supplier capabilities (including ingredients, components, products, services and other capabilities – see also Sect. 3.2). Generally, the more customized and tailored (and hence innovative) the usage of supplier capabilities in relation to particular customer business challenges is, the more difficult it will be for a customer to change to an alternative source of supply. Hence the higher the perceived risk may be. Besides the actual risk, perceived financial and relational switching costs may play a role. Starting point for lowering perceived risk is recognizing, understanding and acknowledging the various sources and forms of perceived risk within the customer organization. Not at a single point in time, but during the entire 'customer activity cycle' (this means before, during and after purchase or at different phases within the relationship if the collaborative relationship is more of an ongoing process). Additionally, rather than trying to anticipate the future, the impact of unforeseen events within customer-supplier relationships may be reduced or even transformed by developing the capabilities and resilience to deal with them in the best possible way (Taleb et al. 2009). Supplier offerings and joint approaches can be elaborated to deal with (unforeseen) risk in the best possible way and increase security and trust in the customer-supplier interaction. This way, the transformative customer-supplier relationship increases in robustness and resilience or even 'antifragility' (Taleb 2012) beyond what would have been possible as seen from the single effort of the separate entities or companies. 'Antifragility is beyond resilience or robustness. The resilience resists shocks and stays the same; the antifragile gets better' (Taleb 2012, p. 3). Superior customer value may not only be realized by trying to avoid or eliminate risk but by additionally challenging customers to change the way they think about risk. Stimulating and guiding them to become in essence more antifragile.

> **Fostering anti-fragility**: superior customer value may not only be realized by trying to avoid or eliminate risk but additionally by challenging customers to change the way they think about risk.

3. Increased positive emotions such as comfort, pride, inspiration and happiness

> To score big with suppliers, you have to win their hearts (Dave Nelson, former Vice President of Purchasing, Honda of America as cited in Blonska et al. 2013, p. 1295).

To build a real connection with major stakeholders requires that positive emotions are experienced within the customer-supplier interaction. The quote above from a customer perspective towards a supplier also applies vice versa. To be able to guide the customer in making their business run better and become more successful, we have to win their hearts first. Emotional impact includes increased positive emotions such as increased comfort, attraction, peace of mind, interest, hope, inspiration, pride or happiness within the customer organization. This starts with having a genuine interest in other people and in our counterparts within the customer organization. Building emotional connections with customers requires 'emotional intelligence' (Goleman 1996). This requires being in touch with one's own emotions first, and additionally being able to understand and relate to the emotions of others, connecting with the customer not only on a rational level but first and foremost on an emotional level.

Goleman (2013) distinguishes between three kinds of empathy that are important for leadership effectiveness. We also believe these are relevant within transformative customer-supplier interactions as well. These are 'cognitive empathy', defined by Goleman (2013, p. 55) as 'the ability to understand another person's perspective', 'emotional empathy', defined as 'the ability to feel what someone else feels' and 'emphatic concern', defined as 'the ability to sense what another person needs from you'. This 'empathy triad' is the basis for increasing positive emotions within customer relations. It is about the understanding what 'moves' people and being able to connect and contribute to the positive emotions experienced. It is not about applying the right techniques or tricks, but about the genuine passion and interest to understand other people and enrich their (working) lives.

Emotional connection: To be able to guide the customer in making their business run better and becoming more successful, we first and foremost have to be able to genuinely connect on an emotional level with prioritized stakeholders.

4. *Increased trust levels*

Trust is a prerequisite to significantly increase competitiveness in all areas of business. Trust in the customer-supplier interaction makes activities more effective and increases efficiency (Castaldo 2007). This may result in a greater joint commitment and willingness to jointly explore and cooperate. Business relations based on trust foster dedication (Chu 2009) and increase 'collective intelligence' (Quinn 2004). At the same time, in a trust-based relationship, control costs are lower and conflicts are solved faster as compared to more transactional relations (Castaldo 2007). The presence (or absence) of trust is a significant driver of economic profit (Maister et al. 2000; Castaldo 2007) and a source of competitive advantage (Peppers and Rogers 2012). As Covey and Merrill (2006) point out, within trust-based relations not only costs decrease, also the speed in which transformation takes place increases significantly. 'When Warren Buffet's Berkshire Hathaway acquired McLane Distribution from Wal-mart, the $ 23 billion acquisition was sealed over a handshake and completed in less than a month, because both parties knew and trusted each other completely. Normally a deal like this would have required six months or more to execute and perhaps several million dollars of legal and accounting fees' (Covey and Merrill 2006, as cited in Peppers and Rogers 2011, p. 81).

> **Trust Equation**
>
> Maister et al. (2000) argue that to generate trust in customer-supplier interactions, it is important to address four elements. These are:
>
> - *Credibility:* based upon the words we speak: to which extent can the other party trust that what you say is true and fact based?
> - *Reliability*: based upon our actions or behavior: to which extent do we do what we promised?
> - *Intimacy*: based upon the closeness of the relationship: to which extent does the other party feel 'safe' with us? How emphatic are we?
> - *Self-orientation*: based upon the experience of the other party of how much we actually care about our own interests or whether we are genuinely interested in the interest of the other. The more the other party experiences that we are focused upon our own interest, the lower the trust within the relation will be. How sincere are we?
>
> Their 'Trust Equation' is built up as follows:
>
> Trust = Credibility + Reliability + Intimacy/Self-orientation.
>
> The Trust Equation indicates that trust levels can be increased by improving each of the four elements. The major impact in increasing trust levels is realized by increasing a genuine focus and orientation towards the interest of the other party, rather than focusing and acting upon one's own (short term) interest solely. This may include advising a strategic customer about (competitive) alternatives if that is in their best interest in specific situations (or including competitive products, services or competencies within the total offer).
>
> Source: Maister et al. (2000)

Exhibit 4.4 Trust Equation (Maister et al. 2000)

To deepen the understanding of existing trust in customer-supplier interactions and to increase trust levels in a genuine way, suppliers may reflect upon the so-called 'trust equation' as provided by Maister et al. (2000).

5. *Decreased total cost of ownership*

Decreasing total cost of ownership (TCO) can be realized by analyzing all direct and indirect costs involved in the suppler-buyer interaction. Examples are the reduction of inventory levels, improvement of capacity utilization, decrease of response times, increase of efficiency in customer-supplier interactions and simplifying ordering, payment and administrative procedures. In fact all costs in the customer-supplier interaction may be analyzed in the stages before, during and after the actual purchase. Ellram (1993, as cited in Axelsson et al. 2005) divides cost components into pre-transaction costs, transaction costs and post-transaction costs. A focus on total cost of ownership may also bring a new perspective on the supplier-buyer collaboration. 'One in which all costs, direct and indirect – in both the customer and supplier – are seen as targets for elimination or significant improvement. The underlying assumption is that these can be improved only jointly, and that it is often one firm's new approach that take cost out of the other' (Cordón and Vollmann 2008, p. 33). This requires transparency and a thorough understanding of the customer's cost structure and a calculation of the monetary worth and impact of supplier competencies and value propositions (see also Sect. 4.3). Festo, for example focuses on increasing the productivity of their global customers by increasing efficiency, reducing energy and reducing hardware and lifecycle costs (see Exhibit 4.5).

6. *Optimized asset utilization*

Next to decreasing the actual costs, cash flow may be improved by changing the asset base or optimizing asset utilization (Axelsson et al. 2005). In this case both the actual costs and the net capital employed may be reduced making it available for other purposes;. for example by outsourcing, smart inventory management or shared use. In 'The Age of Access', Rifkin (2000) describes the ongoing trend from 'owning' to 'using'. 'In the new era, markets are making way for networks, and ownership is steadily being replaced by access (…) Companies around the world are selling off their real estate, shrinking their inventories, leasing their equipment, and outsourcing their activities' (Rifkin 2000, p. 4–5). Ownership of physical property, once considered a value asset, is now regarded as a liability in the corporate world. Customers may increasingly pay for the experience of using things and services, for example by subscriptions, memberships, short-term leases and rentals. Also inventories are reduced or almost disappearing. In order to support customers in optimizing their asset utilization, suppliers may adapt their offering from selling goods and services into making the required competencies and resources accessible to customers upon request. By renting or leasing, they enable or facilitate shared use among different customers or 'access upon demand' (e. g. 'flexcars' and cloud computing) or 'open source' offerings. Instead of selling tires to transportation companies, tire manufacturers offer access to relevant components and services required per mile driven. 'They contract with fleet owners to charge per mile of usage. The pricing contract will be based on the type of use, influenced by general factors such as the type of loads (for example, heavy loads), typical route structures (for example through cities or across long dis-

Festo: Reducing Total Cost of Ownership for Their Global Customers

Festo is a leading world-wide supplier of automation technology and the performance leader in industrial training and education programs. At Festo pursue a common goal: the maximum productivity and competitiveness of their customers in factory and process automation. Festo develops pneumatic and electrical drive and control technology for factory and process automation – from individual catalogue products to ready-to-install systems.

Folkert Hettinga (Industrial Sales Manager Food & Beverage, Agriculture) explained: 'Festo focuses on increasing customer productivity by increasing efficiency, reducing energy and reducing hardware and lifecycle costs. This results in significant energy and lifecycle cost reduction in their customers' processes. For example the leakage of compressed air installations was decreased from 25 to 5 %. For huge production plants like Unilever, Mars, Johnson, etc. the estimate cost savings are between € 50,000 and € 100,000. For such engineering services Festo was awarded from one of their most important customers last year. Festo is a critical part of their customers' team and a true competitive advantage for them'.

Collaborative teamwork and lifecycle analysis to improve machine productivity
Reducing Total Cost of Ownership at global customers is teamwork and requires detailed knowledge sharing and an analysis of the entire value chain in which these customers are active. This goes beyond the traditional seller-buyer relationships and requires true collaboration between business partners. In order to significantly reduce TCO, 'SMART logistics' are applied through the entire lifecycle. Joint performance indicators are defined in relation to Quality, Logistics, Technology and Costs (QLTC indicators – a model originally developed by ASML). From the machine concept development (joint-open-innovation) and prototyping phase (for example applying 3D printing for prototypes) towards installment (offering 'embedded engineers' onsite), utilization (adding remote diagnostics), maintenance and repair (progressive maintenance and reduction of downtime) up to removal. In order to guarantee 'up scaling ' of successful machine productivity lifecycle concepts, a joint innovation road map has been elaborated and implemented.

Information Platform provides easy access to global customers with many plants and engineering centers
Festo ensures that everything they do with their most important customers is transparent. Festo sets up an information platform ('My Festo') that gives all relevant people of their customers quick and central access to different engineering tools and technology neutral comparisons of operating costs, energy costs and product lifecycles. These platforms can also include an online customized catalogue and shop account including an overview of realized projects with Festo anywhere in the world and a list of preferred products and available tools. These tools include also for example CAD drawings, didactic learning tools and contact lists.

Source: Interview with Folkert Hettinga (Industrial Sales Manager Food & Beverage, Agriculture at Festo), April 2014

Exhibit 4.5 Festo: Reducing Total Cost of Ownership for their Global Customers (based on interview with Folkert Hettinga (Industrial Sales Manager Food & Beverage, Agriculture at Festo), April 2014, published with permission)

> **GE Aviation: Increasing 'Residual Value' for Boeing Business Jet Customers**
>
> GE Aviation, an operating unit of General Electric is a world-leading provider of jet, turboprop and turbo shaft engines, components and integrated systems for commercial, military, business and general aviation aircraft. GE Aviation offers their customers a full range of financial and asset management services that help customers meet their business objectives. These services include: 'per hour' engine maintenance, short-term rentals, operational leases (including sales/lease back), structured long term finance solutions and programs increasing 'residual value' of the engines.
>
> Boeing Business Jets has named GE Aviation as the certified engine service company for GE- and CFM56-powered Boeing business jets (BBJs). GE offers long-term 'OnPoint' solution agreements for engine maintenance, repair and overhaul to BBJ customers of new or used aircrafts. These 'On Point' hourly agreements provide comprehensive coverage with predictable maintenance costs and OEM-quality maintenance and parts. The program does not only offer benefits to customers during operation, but also in the resale market. According to aircraft valuation companies, there is a significant increase in the 'residual value' of the aircraft engines. For example the residual value of the CF34-3-powered aircraft covered with an 'OnPoint' agreement is increased by $ 2 million.
>
> Source: GE Aviation (October 2013)

Exhibit 4.6 GE Aviation: increasing 'residual value' for Boeing Business Jet customers (GE 2013)

tances), and individual characteristics of fleet owners, such as the training of the drivers and therefore the quality of driving, the maintenance of correct tire pressure, and the quality of servicing, such as tire rotation. The tire as a product still exists and is at the core of the business. However, the revenue is based on tire usage, not on a one-time tire sale' (Prahalad and Krishnan 2008, p. 15). Another example is a packaging supplier (Greif Incorporated), who, instead of selling drums and containers, 'leases their customers the entire trip', assembling an end-to-end solution including shipment, storage, filling, transportation, cleaning and recycling of the containers (Anderson and Narus 2004). GE Aviation offers their customers a full range of financial and asset management services (see Exhibit 4.6).

7. Increased revenue generating capacity

Suppliers may increase the customers' revenue generating capacity, for example by sharing knowledge and experience about new markets customers are planning to enter. By using or applying the supplier market or technological knowledge, legal insights or established network, the customer may be able to build new value propositions to their customers and grow faster in a new

> **LSI Logic Corporation and VLSI Technology: Enabling Customer's Customization**
>
> 'Computer Chip producers like LSI Logic Corporation and VLSI Technology provide their business customers with do-it-yourself tools that enable customer-chip-based manufactures (for example toy manufacturers that need circuitry in their products) to design their own specialized chips, thus taking the customized customer chip market from virtually nothing to more than $ 20 billion'.
>
> Source: O'Cass and Ngo (2012, p. 133)

Exhibit 4.7 LSI Logic Corporation and VLSI Technology: enabling customer's customization (O'Cass and Ngo 2012, p. 133)

market. Customers and suppliers may also jointly analyze ways to address changing demands in downstream customer markets. Revenue generating capacity may furthermore be increased by sharing insights or technologies, facilitating the customer to build a first mover advantage within their market or redefine standards within the customer's industry. Another interesting approach is to provide tools or methods to customers to facilitate a better customization of their products and services towards their downstream customers (as illustrated by the example of LSI Logic Corporation and VLSI Technology in Exhibit 4.7).

Additionally, customers may be supported in increasing customer satisfaction in their markets by increasing the satisfaction of their downstream customers. Support may be given to increase responsiveness to customer's customers, speeding up 'time to market,' or by the early development or innovation of new (or improved) products and services better aligned with changing customer's customers demand.

Furthermore, in a joint effort, the customer's competitive advantage may be increased by increasing the customer's customers' experience. Vodafone and Amazon for example work together to increase Amazon's customers' experience (see Exhibit 4.9).

8. *Increased revenues*

Influenced by the traditional focus of purchasers on reducing costs (or reducing prices), a less explored area in customer-supplier relations is the potential to increase the customer's competitiveness by increasing revenue growth. Examples would include connecting customers to potential customers or distribution partners in new markets or by using the company network or relational capital to connect customers to (joint) new business opportunities. There may

Kodak: Accelerating Time-to-Market for Consumer Goods Producing Companies

'For consumer goods product companies, accelerating the time to market for new products is a strategic imperative. Often the packaging design and development process appear to be a bottleneck for achieving this objective. In reality the challenge exists in the numerous activities that provide label content and information – including marketing promotions, product specifications, artwork, ingredients labeling, compliance, bar-coding and branding – which all come together during package design and development. Typically, these upstream processes fall outside the control of the packaging function. Kodak accelerates the time-to-market of their consumer goods producing customers by redefining the ways these functions interact with packaging. By reducing the time-to-market of a product, consumer goods producing companies can accelerate the revenue realization and reduce product and package development costs'.

Source: Industry Insights From Kodak, Accelerating Product Time-to-Market, Kodak, 2009.

Exhibit 4.8 Kodak: accelerating time-to-market for consumer goods producing companies (Kodak 2009)

Joint Go-to-Market: Vodafone and Amazon to Increase 'Always on Experience'

Vodafone Group Plc is one of the world's leading telecommunications groups, with a significant presence in Europe, the Middle East, Africa and Asia Pacific. The company's group revenue amounts € 57.3 billion (year ended 31 March 2014). Vodafone supports its global customers in creating 'a competitive edge' using its global technology and networks. The goal of its enterprise customer Amazon is 'to be the earth's most customer centric company for four primary customer groups: consumers, resellers, enterprises and content creators'. Operating 24/7 in five continents, to customers who expect an always-on experience, there is no downtime. Customers increasingly want to buy, download and read digitized content from any device, traveling anywhere. For this reason Amazon designed and built its own eReader. In order to allow users easy access to content anytime and to ensure a smooth launch, Amazon needed a connectivity partner capable of delivering a consistent global user experience. In a joint-go-to-market between Vodafone and Amazon the first Kindle (eReader) with 3G connectivity was launched in the US in 2010 and went global in 2011. Today Kindle paperwhite 3G users can download a book in more than 150 countries. The success of the Kindle has paved the way for more ambitious hardware. It was also the platform upon which the Amazon and Vodafone relationship grew from a supplier-buyer conversation into a more connected partnership.

Source: Vodafone (2014) – Vodafone Global Enterprise Amazon Case Study (published on Vodafone website)

Exhibit 4.9 Joint Go-to-Market: Vodafone and Amazon to increase 'always on experience' (Source: Vodafone 2014 – Vodafone Global Enterprise Amazon Case study, published on Vodafone website; reproduced with permission)

also be potential at the supplier organization to increase reciprocal value. Marriott invited strategic customer Siemens to bring the latest web technology to business travelers staying at Marriott hotels. This resulted not only in increased revenue growth for Siemens at Marriott; it also resulted in a Marriott-Siemens high-tech joint venture offering web solutions to other hotels (Sherman et al. 2003). Also 'UPS and Kodak have established reciprocal relationships where UPS gains value from the relationship in the area of Kodak's imaging expertise and Kodak benefits from UPS distribution capabilities' (O'Cass and Ngo 2012, p. 133).

4.2 Acting as a Lead Collaborator in Global Value Networks

> Do what you do best and link to the rest (Jarvis 2009, p. 26).

In order to mobilize the required competencies and resources to guide customers in becoming more competitive and successful in their markets requires a shift in focus from traditional, linear value chains to becoming a 'lead collaborator' (Vitale et al. 2011) in nonlinear, global value networks. To be able to address the customer business challenges and headaches may require value innovation beyond existing products and services, tapping into both resources and competencies in the supplier organization and at accessible reach within the value network. 'These parties, or collaborators, can either be companies that the [supplier] previously allied with or may be a competitor in another market. Thus the chain expands into a multidimensional network. This network of collaborators includes nontraditional partners in a way that all partners in the network 'win' as part of the team that provides the offering of greatest value' (Vitale et al. 2011, p. 42).

To make this happen, suppliers (or their strategic customers) may act as 'lead collaborators' (Vitale et al. 2011) in the value network. These value networks may be composed in different ways, depending upon the chosen priorities on the joint transformation agenda. Within ICT industries this change from linear value chains to nonlinear value networks is clearly visible. Instead of a chain of parties (e. g. content providers delivering to aggregators, delivering to application & platform and service providers, delivering to end-users), value networks are in many cases formed around strategic customers (see for example Fig. 4.3 depicting an example of a value network within Lucent-Alcatel).

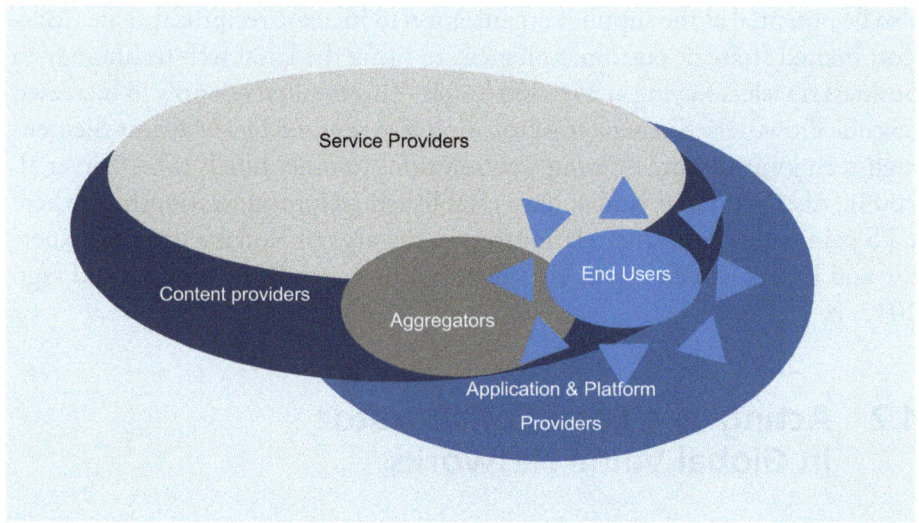

Fig. 4.3 Moving to nonlinear value networks: example ICT

Ritter and Gemünden (2003) present empirical evidence that 'network competence' impacts innovation success and expect network competence to have an impact on the performance of supply networks as well. This idea is supported by Prahalad and Krishnan (2008) who observe new ways of innovation and co-creation through collaboration within the global value network between customers, suppliers and channel partners interconnected in 'global innovation networks'. As argued earlier by Prahalad and Ramaswamy (2004), we are moving towards 'robust experience networks' In these value networks competences reside in an 'enhanced network'. In order to co-create value competencies of firms within the network, these must be 'selectively activated to create unique value', (Prahalad and Ramaswamy 2004, p. 116).

As argued by Theo Verweerden (Marketing Program Director Value Creation, Royal DSM): 'Value innovation requires the inclusion of partners within and across value chains, in other words we need to include the entire 'value eco system'. Within DSM we have numerous examples of creating value by tapping into the broad eco system. Within our Engineering Plastics business (global supplier of high-performance engineering thermoplastic solutions), this way of innovating with several partners in our value eco system is the only way forward to drive sustainability and to come up with innovative solutions. Recently this type of collaboration resulted in a complete new lightweight earset solution, part of which (new cable type) included by DSM'.

To activate the competencies and resources required, traditional company borders disappear. As suggested by Ridderstråle and Nordström (2004), com-

> **Procter and Gamble: Joining Forces with Competitors to Improve Supply Chain Efficiency of Strategic Retail Customers**
>
> 'Facing the need for more frequent low volume deliveries into store, the adoption of everyday low pricing (EDLP), the requirement for suppliers to take a proactive role in managing categories of products, including those supplied by competitors, and the opportunities provided by e-business, P&G has joined forces with its competitors to form *Transora* a major FMCG Web portal designed to facilitate more collaborative use of logistics resources and improve the retail industry's supply chain efficiency'.
>
> Source: Wilson et al. (2001, p. 73)

Exhibit 4.10 Procter and Gamble: joining forces with competitors to improve supply chain efficiency of retailers (Wilson et al. 2001, p. 73)

panies within the value network could resemble a 'Lego-model', building cross boundary networks, reinventing value by connecting 'bricks' in new meaningful ways. Value in these networks can be created moving beyond traditional company borders and value chains 'taking [the original companies] apart and moving the pieces [competencies] around' (Ridderstråle and Nordström 2004, p. 131). In fact in many cases the only way to disrupt the current way of doing business and increase value substantially is to establish new forms of collaboration with partners, including the competition in the value network.

4.3 Business-Altering Value Propositions

Transformational sales is about guiding change and learning, both for the customer and for one's own organization (see also Chap. 5). Fundamental to inspiring change is the ability to envision the joint future we seek to create with our customers and to express this in a way that 'moves' people. This means to vocalize (and if possible visualize) the essence of what the increased competitiveness will look like, touching upon the untapped business potential within the customer organization and personalizing the impact for key stakeholders. In other words it is vital to craft and communicate value propositions in a way that customers feel and understand. As Anderson et al. (2006) phrase it, to craft value propositions with 'resonating focus'.

Even though the term value proposition is often used in a business context, in practice not many propositions have 'resonating impact' to customers. Payne and Frow (2014), referring to their recent study (Frow and Payne 2012) mention that out of over 200 companies surveyed, less than 10 per cent formally develop effective value propositions. As commented by Anderson et al.

Type of proposition	All Benefits	Favorable points of difference (USP's)	'Resonating Impact'
Consists of	All benefits customers receive from a market offering	All favorable points of difference a market offering has relative to the next best alternative	The 1 or 2 most relevant issues whose improvement will make the greatest impact to the customer business (UBR)
Purpose	Convince	Differentiate	Make a difference

Fig. 4.4 Three ways in which value propositions are conveyed

(2006, p. 93), in practice companies adopt three alternative ways to convey value to their customers:

- *All benefits*: focuses on all benefits customers may receive from the supplier offering. In fact in this case the supplier tries to *convince* the customer why they should buy from them.
- *Favorable points of difference*: focuses on the major points of differences (or unique selling points) the supplier may offer as compared to the alternatives considered by the customer. In fact in this case the supplier tries to *differentiate* themselves in the best possible way, as compared to others.
- *Resonating focus*: focuses on what is most relevant for customers, 'in a way that conveys a sophisticated understanding of the customer's business priorities' and answers the question what is '*most worthwhile*' for the customer to keep in mind about the offering. We could say the essence of this type of proposition is to *make a difference* to the customer business (see also Kotler et al. 2010).

Based on Anderson et al. (2006) we would like to depict the three types of value propositions as shown in Fig. 4.4.

Obviously, value propositions with 'resonating focus' have the largest impact. Part of guiding customer business transformation is conveying the value offered in a resonating way and touching upon the customer's untapped business potential to make a real difference for and with them. Building on the model of Anderson et al. (2006) we would say a resonating business-altering value proposition contains the following elements (see Fig. 4.5).

Making a difference…	
Rational and Emotional Connection	• Convey impact in a way the customer feels and understands • Touch upon **untapped customer business potential** • Use customer business language and 'money-making logic'
Relevance: Less is more	• Articulate the 1 or 2 most relevant issues whose improvement • will make **the greatest impact to customer success in their markets**
Customer Insight	• Broaden the customer *perspective* on their way of doing business • and on the *change in assumptions, behavior, processes and collaboration* required • to release the previously untapped business potential
Supplier Impact	• Demonstrate the **business-altering contribution** in monetary terms • Personalize to targeted stakeholders • Bring to life through genuine connection with personal professional mission

Fig. 4.5 Business-altering value propositions

Rational and Emotional Connection

To increase resonance, value propositions have to connect and touch upon what really matters to the customer. To connect in this way, the possible impact or business-altering contribution has to be conveyed and expressed in a way the customer feels and understands, connecting not only on a rational level, but first and foremost on an emotional level to what is really important and relevant to them and thereby touching upon the untapped customer business potential. This means enabling the customer to see the business potential that is there to be released, but which is previously unseen. At the same time they are inspiring and uplifting stakeholders within the customer organization, and also inspire them to learn and grow. We believe that bringing a value proposition to life requires a genuine interest in the other. To engage with the customer in an open way and to make a difference requires being 'other focused' and at the same time connected to an inner purpose or as referred to by Quinn (2004) to be 'purpose centered'. We would say this requires a genuine connection to a deeper purpose or professional mission of the strategic salespeople involved (see also Chap. 6).

> **To bring a value proposition to life** requires enabling the customer to see what is there, but which is unseen and to touch upon the business potential that is there but which is untapped. Foremost this requires a genuine connection to a deeper purpose of the strategic salespeople involved. A genuine desire to make a difference.

Vocalize in Customer Business Language and Money Making Logic

To increase resonance a value proposition is preferably crafted in the 'business language' that is used within the customer organization. This permits to connect to the customer's strategic focus and value drivers by using the language and terminology that are most common within the customer organization rather than choosing the 'jargon' that is around and well known within the supplier organization. Preferably the contribution to the customer business is calculated and vocalized in the 'moneymaking logic' (e. g. Cheverton 2008) of the customer. If an airline calculates results in 'profit per seat' (their money making logic), the value proposition of a strategic supplier will resonate stronger (and has a larger impact) if it articulates the contribution in terms of 'profit per seat' as well. Whereas for poultry producers making business decisions on a 'cost per chicken basis', the impact may be defined in terms of a 'decreased cost per chicken'. For large retailers like Wal-Mart and Tesco with relatively low margins and a high capital turnover ratio, contribution to their *velocity* is vital. As argued by Cheverton (2008): 'Wal-Mart and Tesco share a similar moneymaking logic. They both need scale, they both need what they call 'velocity' (high stock turn, and the healthy cash flow that should come with it), and they are both happy to work on low margins to ensure they get the scale and the velocity required'.

Focus Upon What is Most Relevant to the Customer

Business-altering value propositions are focused on what is most relevant to the customer. They elicit profound Customer Insight. They articulate and focus on the customer's major reason to be engaged with the supplier. As argued by Anderson et al. (2006), they convey the one or two most relevant issues whose improvement will make the greatest impact to the customer business. These are the business altering elements that will make the customer more successful and competitive in their markets. In addition to the perspective on the possible business impact, it is important to elicit insight into the change

> **Value-Bridge at TNT: Design a Close to Damage Free Process**
>
> TNT Express is a global company operating in 200 countries around the world. As Hugo Koppelaars (Director Sales, TNT Germany) says: 'Creating customer value requires a thorough understanding of the customer and their business processes. We developed models with which we can demonstrate the true value of our service compared to our competitors. These models translate our product features, like speed and on-time performance, into clear customer benefits, like reduced inventory and lower service costs. The output is what we call a 'value bridge' in which we show the true cost of using TNT (price minus unique value) compared to the best alternative. To create and fill these models it is crucial for an Account manager to understand the impact of logistics in the customer process. We calculated for one of our customers that a decrease of 0.01 percent in losses during transport will result in €167,000,– savings per year on their side. Express providers usually focus on the speed and reliability of delivery, but in this case it was much more interesting to design a close to damage free process. This resulted in a process in which losses during transport were minimized'.
>
> Source: Interview with Hugo Koppelaars (Director Sales TNT), February 2013

Exhibit 4.11 Value-bridge at TNT: design a close to damage free process (based on interview with Hugo Koppelaars, Director Sales TNT, February 2013; published with permission)

in assumptions, behavior and collaboration required to realize the business transformation.

Use a Value Bridge to Demonstrate the Actual Business-Altering and Monetary Impact

Preferably both the business-altering aspect (*which part of the customer business or behavior is changed in interaction with the supplier?*) and the (monetary) impact for the customer are highlighted and related to the prioritized customer business challenges. To demonstrate the monetary impact, a resonating value proposition includes a 'value word equation' (Anderson et al. 2008) or 'value bridge', documenting the actual value provided to a customer, preferably in financial terms.

As argued by Anderson et al. (2008, p. 52) 'value word equations provide a methodical way of convincingly demonstrating and documenting superior value in monetary terms'. They provide an example of Rockwell Automation that used to calculate the cost savings for their customers based upon the reduced power usage that a customer would gain by using a Rockwell pump as compared to alternative solutions. Even if in practice it is not always possible to calculate the financial value of an offering, it is worthwhile to make the 'size of the value opportunity' visible to the customer (Terho et al. 2012). To additionally increase the resonance of the proposition, it is important to tailor the conversation to the individual stakeholders within the customer organization.

Tailoring the Conversation: Personalized Value Propositions

In addition to the generic value proposition, personalized propositions need to be elaborated to address the different individuals within the customer's *decision making unit*, highlighting these elements that are most worthwhile for each stakeholder, tailoring the conversation in such a way that specific challenges of the key stakeholders in the customer buying center are addressed and recognizing that there is untapped potential in every person. This provides a new perspective upon the possible impact that may be realized and the (behavioral) change required. A personalized business-altering value proposition should contain at least the following elements.

- *Person*: a selected key stakeholder within the customer organization to be addressed (starting with selected 'change agents' – open to change).
- *Potential: prioritized customer business challenge or headache:* one that is most relevant to the particular stakeholder (what do they care most about?) and that contains *untapped business potential*.
- *Perspective*: envisioning what the new situation will look like (*possible impact: increased customer success*) and how this will be realized (*mental and behavioral change required*).
- *Reason to believe:* supportive arguments (evidence) that demonstrate why the supplier is equipped to guide the proposed transformation (proven impact).

For an ICT-supplier doing business with strategic customers in the healthcare industry, a personalized value proposition may be built up as shown in Exhibit 4.12.

Reciprocal Value: Demonstrating Win-Win

Beyond calculating and demonstrating the value of the customer-supplier interaction to the customer, transformational sales requires calculating and demonstrating the value of the chosen business ventures with the customer. As argued by Walter et al. (2001, p. 373) 'given that customers today expect to be involved in relationships in order to gain benefits of their own, suppliers need to understand the potential which these relationships offer them in return' (see also Sect. 5.2 on joint Profit and Loss statements). Ballantyne et al. (2011) suggest building reciprocal value propositions which clarify what

> **Value Proposition of ICT-Supplier to Strategic Healthcare Customer**
> - 'For hospital group management (*selected stakeholder in customer buying center*)
> - facing the challenge of increased quality transparency (*most relevant customer business challenge with untapped potential*)
> - ICT supplier X provides a radical new perspective on the interaction between patients and healthcare professionals (*perspective upon business change*)
> - resulting in a 15 percent reduction of errors made within your hospital group (*perspective upon business impact*)
> - because of real-time patient information accessible to all healthcare professionals and patients, which has resulted in an average decrease of errors made of more than 15 percent in 50 pilot hospitals worldwide (*proven impact* of distinctive capabilities)'

Exhibit 4.12 Value proposition of ICT-supplier to strategic healthcare customer

is expected and gained by both parties. This might be a useful approach to articulate and demonstrate win-win.

Win-Win

> Win-win is not hitting the other person twice (Cordón and Vollmann 2008, p. 21).

In true collaborative relationships, both parties have a shared and genuine interest in the impact of the collaboration for both parties. However, even in collaborative relationships, with a shared transformation agenda and joint long-term objectives, conflicting interests may exist in the customer-supplier relationship. To creatively explore possibilities to deal with potentially conflicting interests in a constructive way requires a genuine interest and willingness from both parties to change perspective and to maximize both one's own interest and that of the other party. In their 'Thomas-Kilmann Conflict mode instrument' (Thomas and Kilmann 2002; see also Tanner 2014) identify different styles in which parties may deal with conflicting interests. They separate two dimensions: the level of cooperation (*focus on the interest of the other*) and the level of assertiveness (*focus on one's own interest*). Based upon the way in which two parties combine both interests, they distinguish five styles: competing (win-lose), avoiding (lose-lose), accommodating (lose-win), compromising (half-half) and collaborating (win-win). Figure 4.6 depicts an adapted version of this model applied to the customer-supplier interaction.

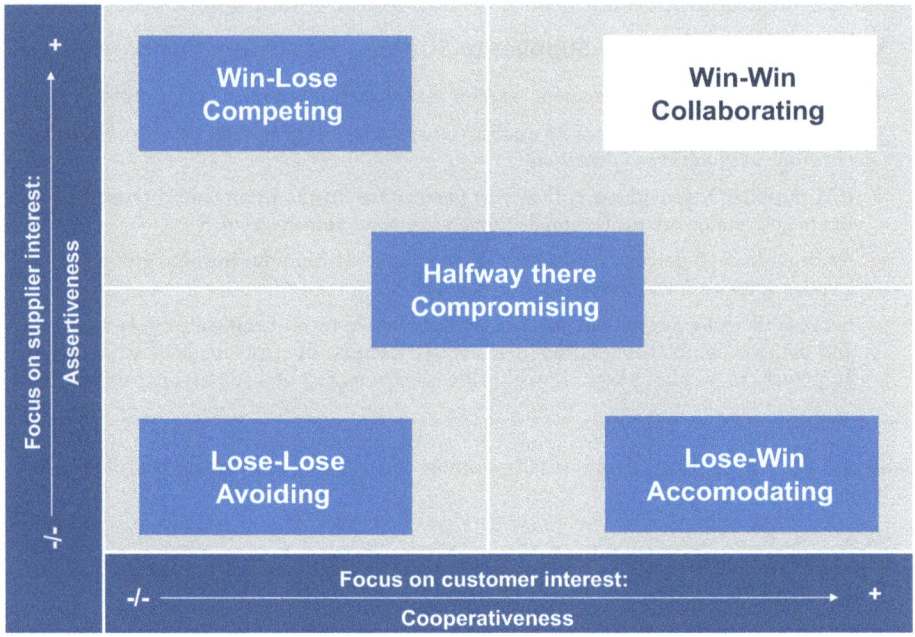

Fig. 4.6 Win-win: Thomas-Kilmann Instrument in customer-supplier interaction (supplier perspective)

In our experience, many parties in customer-supplier interactions settle halfway. We would argue that particularly in transformative customer-supplier interactions, it is important to move beyond compromise and explore possibilities to create a true win-win for both parties. This may require a longer joint time horizon and long term planning in joint innovation and developing the customer's redefined (future) products. ABB and Caterpillar for example are jointly developing next-generation diesel engines. ABB (as a manufacturer of turbochargers) can help Caterpillar to achieve increased engine performance while at the same time reducing pollution. In addition ABB technical competencies in combustion, sensing and emissions control may bring potential value to Caterpillar in its development of next-generation diesel engines. 'The task before this pair of firms is how to find the best mode of collaboration to develop the true win-win' (Cordón and Vollmann 2008, p. 38).

Besides longer-term joint objectives, win-wins can be defined for new upcoming joint projects. Clarifying win-wins will also enable internal transformation. In addition to building resonating business-altering value propositions to convey the value-innovation to the external customers, transformational sales requires to build resonating propositions to the internal organization as well. A resonating proposition will not only contribute to inspiring change within the customer organization, it will also be useful to inspire change within one's own organization (see Chap. 5 on internal transformation).

References

Anderson, J. C., & Narus, J. A. (2004). *Business Market Management. Understanding, Creating and Delivering Value*. Upper Saddle River: Pearson, Prentice Hall.

Anderson, J. C., Narus, J. A., & van Rossum, W. (2006). Customer Value Propositions in Business Markets. *Harvard Business Review, 84*(3), 90–99.

Anderson, J. C., Kumar, N., & Narus, J. A. (2008). Certified Value Sellers. *Business Strategy Review, 19*(1), 48–53.

Ateş, M. A. (2014). *Purchasing and Supply Management at the Purchase Category Level: strategy, structure and performance*. Rotterdam: Erasmus Research Institute of Management (ERIM).

Axelsson, B., Rozemeijer, F., & Wynstra, F. (2005). *Developing Sourcing Capabilities*. Chichester: John Wiley & Sons, Ltd.

Ballantyne, D., Frow, P., Varey, R. J., & Payne, A. (2011). Value Propositions as Communication Practice: Taking a Wider View. *Industrial Marketing Management, 40*(2), 202–210.

Biemans, W. G. (2010). *Business to business marketing: a value-driven approach*. Maidenhead: McGraw-Hill Higher Education.

Blonska, A., Storey, C., Rozemeijer, F., Wetzels, M., & de Ruyter, K. (2013). Decomposing the effect of supplier development on relationship benefits: the role of relational capital. *Industrial Marketing Management, 42*, 1295–1306.

Castaldo, S. (2007). *Trust in Market Relationships*. Cheltenham: Edward Elgar Publishing.

Cheverton, P. (2008). *Global Account Management*. London: Kogan Page.

Chu, K. (2009). The construction model of customer trust, perceived value and customer loyalty. *The Journal of American Academy of Business, 14*(2 (March)), 98–103.

Cordón, C., & Vollmann, T. E. (2008). *The Power of Two. How Smart Companies Create Win-Win Customer-Supplier Partnerships That Outperform the Competition*. Basingstoke: Palgrave MacMillan.

Covey, S. M. R., & Merrill, R. (2006). *The Speed of Trust: The one thing that changes everything*. New York: Free Press.

Europcar Activity Report, 2011–2012.

Frow, P. and A. Payne, 2012, 'Diagnosing the value proposition', working paper, Discipline of Marketing, University of Sydney, Sydney.

GE, 2013, 'GE enhances open solutions for business jet operators', GE Aviation, October 21, 2013.

Goleman, D. (1996). *Emotional Intelligence*. London: Bloomsbury.

Goleman, D. (2013). The Focused Leader. *Harvard Business Review, 91*(12), 51–60.

Grönroos, C. (2007). *Service Management and Marketing. Customer Management in Service Competition*. Chichester: John Wiley & Sons.

Grönroos, C. (2011). A Service Perspective on Business Relationships: The Value Creation, Interaction and Marketing Interface. *Industrial Marketing Management, 40*(2), 240–247.

Grönroos, C., & Voima, P. (2013). Critical Service Logic: making sense of value creation and co-creation. *Journal of the Academy of Marketing Science, 41*, 133–150.

Jarvis, J. (2009). *What would Google Do?* New York: Harper Colins.

Kodak, 2009, Industry Insights from Kodak. Accelerating Product Time-to Market.

Kotler, P., Kartajaya, H., & Setiawan, I. (2010). *Marketing 3.0*. Hoboken: John Wiley & Sons.

Maister, D. H., Green, C. H., & Galford, R. M. (2000). *The Trusted Advisor*. New York: The Free Press.

Moss Kanter, R. (2008). Transforming Giants. What kind of company makes it its business to make the world a better place? *Harvard Business Review*, (January), 43–52.

O'Cass, A., & Ngo, L. V. (2012). Creating Superior Customer Value for B2B Firms through Supplier Firm Capabilities. *Industrial Marketing Management, 41*(1), 125–135.

Payne, A., & Frow, P. (2014). Developing superior value propositions: a strategic marketing imperative. *Journal of Service Management, 25*(2), 213–227.

Peppers, D., & Rogers, M. (2011). *Managing Customer Relationships: A Strategic Framework*. Hoboken: Wiley.

Peppers, D., & Rogers, M. (2012). *Extreme Trust: Honesty as a Competitive Advantage*. New York: Penguin/Portfolio.

Pine, B. J., & Gilmore, J. H. (2014). A Leader's Guide to Innovation in the Experience Economy. *Strategy &amp Leadership, 42*(1), 24–29.

Prahalad, C. K., & Krishnan, M. S. (2008). *The New Age of Innovation: Driving Co-Created Value through Global Networks*. New York: McGraw Hill.

Prahalad, C. K., & Ramaswamy, V. (2004). Co-Creation Experiences: The Next Practice in Value Creation. *Journal of Interactive Marketing, 18*(3), 5–14.

Quinn, R. E. (2004). *Building The Bridge as you walk on it*. San Francisco: John Wiley & Sons.

Ridderstråle, J., & Nordström, K. (2004). *Karaoke Capitalism, Financial Times.* Stockholm: Prentice Hall.

Rifkin, J. (2000). *The Age of Access. The New Culture of Hypercapitalism where all life is a Paid for experience.* New York: Jeremy P. Tarcher, Putnam, a member of Penguin Putnam Inc..

Ritter, T., & Gemünden, H. G. (2003). Network Competence: Its Impact on Innovation Success and its Antecedents. *Journal of Business Research, 56*(9), 745–755.

Sánchez-Fernández, R., & Iniesta-Bonillo, M. A. (2007). The concept of perceived value: a systematic review of the research. *Marketing Theory, 7*(4), 427–451.

Sherman, S., Sperry, J., & Reese, S. (2003). *The Seven Keys to Managing Strategic Accounts.* New York: McGraw-Hill.

Taleb, N. N. (2012). *Antifragile.* London: Penguin Books.

Taleb, N. N., Goldstein, D. G., & Spitznagel, M. W. (2009). The Six Mistakes Executives make in risk management. *Harvard Business Review, 87*(10), 78–81.

Tanner, R. 2014, *Organizational conflict: Get used to it and use it.* May, 13, 2014, Amazon Digital Services.

Terho, H., Haas, A., Eggert, A., & Ulaga, W. (2012). It's Almost Like Taking the Sales Out of Selling – Towards a Conceptualization of Value-Based Selling in Business Markets. *Industrial Marketing Management, 41*(1), 174–185.

Thomas, K. W., & Kilmann, R. (2002). *Thomas-Kilmann Conflict Mode Instrument.* Mountain View: CPP Inc..

Vitale, R., Giglierano, J., & Pfoertsch, W. (2011). *Business-to-Business Marketing: Analysis and Practice.* Upper Saddle River: Pearson Education.

Vodafone, 2014, Vodafone Global Enterprise. Amazon Case Study, published on Vodafone website.

Walter, A., Ritter, T., & Gemünden, H. G. (2001). Value Creation in Buyer-Seller Relationships. *Industrial Marketing Management, 30*(4), 365–377.

Wilson, K., Millman, T., Weilbaker, D., & Croom, S. (2001). *Harnessing Global Potential. Insights into Managing Customers Worldwide.* Chicago: Strategic Account Management Association.

Zeithaml, V. A. (1988). Consumer Perceptions of Price, Quality and Value: A Means-end model and Synthesis of Evidence. *Journal of Marketing, 52*(3), 2–22.

5
Enabling Internal Transformation

*Our best Account Managers sell their ideas with equal
skill and attention internally and externally*
                                  Hugo Koppelaars, Director Sales TNT

Transformational sales is not limited to strategizing with the external customer; it also requires internal sales and network building. In order to mobilize the required resources to initiate new business ventures with strategic customers and to actually increase twain customer and supplier competitiveness, strategic sales teams need to influence and inspire both external and internal customers. Transformational sales requires disrupting both the customer's and the supplier's thinking and assumptions about their business. In practice, alignment is required among all stakeholders involved. In a recent study among over 100 strategic account managers (Dingena and Teven 2015) 'influencing internally, creating alignment and mobilizing resources' was mentioned as one of the major challenges in strategic sales practice (see Exhibit 5.1).

We believe that one of the crucial elements in why influencing internally is seen as such a major challenge is the underlying assumption in sales that selling is (or should be) an 'outside job'. Overcoming this challenge often requires a change of mind in the first instance.

> **Enabling internal transformation requires first of all a change of mind**
>
> Vital to the success of strategic sales teams is the (change in) mindset. Establishing internal relations is of equal importance as building external relations *and it could be lots of fun!*

As compared to transactional and solution selling approaches, transformational sales requires an *increased internal focus* and network building. Speakman and Ryals (2012) even refer to strategic sales as 'the inside selling job'. True value transformation can only be enabled once 'internal customers' have been identified and internal relations have been built (see Sect. 5.1). In order to align stakeholders around customer business challenges and to mobilize

> **Top 7 Challenges in Strategic Sales Practice**
> **(n > 100 Strategic Account Managers)**
>
> 1. Influencing internally; creating alignment and mobilizing resources
> 2. Becoming a true strategic business partner
> 3. Conducting multi-level relations, including the C-level engagement
> 4. Aligning purchasing and sales perspectives: dealing with purchasers focusing on price reductions rather than cost reduction or revenue increase
> 5. Performance measurement and management
> 6. Business, Market and Customer Intelligence
> 7. Implementation of strategies in practice
>
> Source: Dingena and Teven (2015)

Exhibit 5.1 Top 7 Challenges in Strategic Sales practice (Dingena and Teven 2015)

the required resources, transformational sales requires an *intra*preneurial role and mindset within strategic sales teams. Starting point is the perception of transformational sales as a business rather than as a sales initiative and creating alignment around solid business cases (see Sect. 5.2)

As argued by Helsing et al. (2003), the leverage of internal resources requires a 'greater impact within the own company'. In order to guide change, strategic sales team members need to consciously decide upon their roles in leading change. Driving change successfully requires strategic sales team members to inspire colleagues to contribute their best to new business ventures with strategic customers. A transformative approach may be to become 'the silent conductor', enabling 'leadership from any chair' (Zander and Zander 2000; see Sect. 5.3).

5.1 Building Your Internal Network: The Inside Selling Role

If, as the African proverb goes, it takes a village to raise a child, it takes an entire firm to own and manage a strategic account relationship (Sherman et al. 2003, p. 36).

Crucial to the success of strategic sales teams is the mindset that building internal relations is of equal importance and fun as building external relations (see Fig. 5.1). As compared to transactional and solution selling approaches,

From Customer Insight to Solid Business Development at TNT

TNT Express is a global company operating in 200 countries around the world. As Martijn Legemaat, (Corporate Account Insight Director TNT) says: 'Gaining innovative Customer Insight requires to recognize customer needs in the stage before they are made explicit by customers. And then take the journey with customers to co-create value. In essence this means to be 'One Step ahead of Customer Demand'. The Corporate Account Management department at TNT serves our largest customers worldwide. We have implemented an entrepreneurial sales approach called **TRACK** (Trends, Relationships, Alignment & Create, Knowledge). TRACK starts from the outside-in with market *Trends* that we derive from industry knowledge, as well as trend sessions organized with our most innovative customers in target industries. To be able to prioritize trends, upcoming trends and their impact on supply chains are discussed, challenged and validated in these sessions. This requires collaboration at senior levels with customers, building strategic *Relationships* to truly understand and validate the impact of upcoming trends.

Our sales teams regularly identify major opportunities which complement our core business but may require development of enhanced or new TNT capabilities. In order to combine both the 'market sensing' and 'business economics' perspectives at an early stage we have created so-called *Sales Cases*. In these cases a solid analysis is made about the potential future profitability of emerging opportunities starting from industry trends. Each Sales Case includes specific criteria to be met in order to proceed and get a formal 'go'. An important indicator is for example 'scalability'. To which extent is the emerging trend relevant for more customers within the same industry and could the new or to be enhanced solutions be leveraged?

Sales and Business Case: World Leading Consumer Electronics Manufacturer
For example the market trend of 'rising online sales' was the main driver for a world leading consumer electronics manufacturer to consider delivering consumer electronics directly to end consumers. This customer expressed the desire to start a multichannel distribution model in Europe. The Sales case included changing market trends leading to this need, pains in the current distribution model, the envisioned set-up, outcomes and detailed requirements of how both the customer and TNT could make money from this opportunity.

Once the Sales Case is approved our TNT Global Solutions team (*Alignment* and *Create*) takes it a step further. This team builds alignment, manages expectations and mobilizes resources from other TNT functions to actually design and co-create new services with customers. Bridging both TNT and Customer capabilities and launching new value propositions. To leverage best practices to the fullest potential our experience and *Knowledge are shared* with other TNT sales channels'.

Source: Interviews with Martijn Legemaat (Corporate Account Insight Director TNT) in the period June 2013–January 2014 (Note: see Sect. 5.2 for the further elaboration of the Alignment and Create phase by the Global Solutions Team)

Exhibit 5.2 From Customer Insight to solid business development at TNT (based on interviews with Martijn Legemaat, Corporate Account Insight Director at TNT, June 2013–January 2014; published with permission)

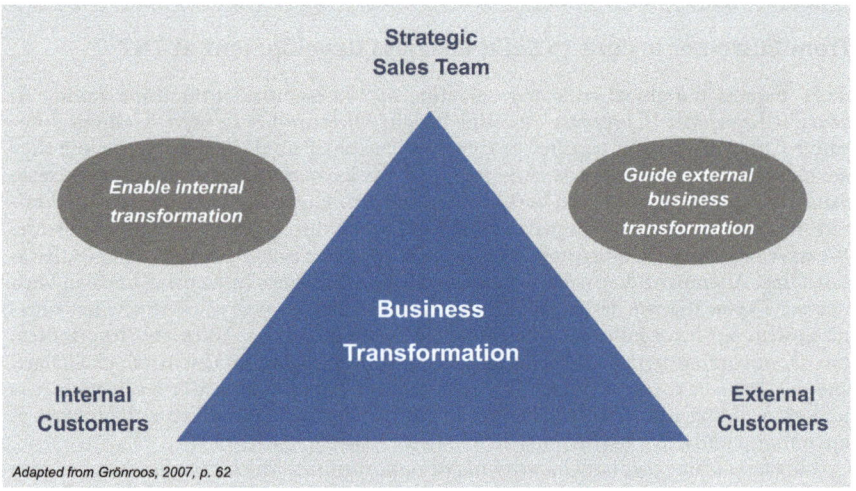

Fig. 5.1 Business transformation requires an integrative perspective on the sales role

transformational sales demands an *increased internal focus* and network building. This requires the ability to connect with internal stakeholders effectively across business units, functions and in many cases across regions within the world. In fact, this means consciously applying the competencies needed to build and maintain relationships with strategic customers externally internally within the own organization towards colleagues.

Based on the consultative selling experience with external customers, most strategic salespeople are very well equipped to build internal relations. Speakman and Ryals (2012) found that salespeople who applied 'adaptive selling practices' also internally, performed better in solving internal conflicts. The reason why internal alignment in practice is found to be a challenge is not so much related to the capacities of strategic salespeople, but rather hampered by the perspective on the sales role. Beyond the capability to build relations, an integrated perspective upon the sales role, and a genuine interest to build relations with all stakeholders involved (both externally and internally) is required, thereby broadening the analysis of the decision making unit and prioritization of stakeholders across company borders.

> **Business transformation requires an integrative perspective on the sales role:** having a genuine interest and passion to identify and build relations with all relevant stakeholders, both within the customer and within the supplier organization. In fact broadening the Decision Making Unit across company borders.

> **Connecting People**
>
> 'What's really important to the customer is how well you understand your own company. It's not so much that the customer expects you to know that this widget has 2.35 units of tolerance. The customer expects you to know who in your company has the ability to solve problems and create customer solutions. So the expectation is more than 'know your product', which is a given, but know your company well enough.'
>
> Part of quote of global account manager as cited in Helsing et al. (2003, p. 53)

Exhibit 5.3 Connecting People (Helsing et al. 2003, p. 53)

Identifying Internal Stakeholders

Being able to build internal networks, starts with understanding one's own company. This includes understanding the basic company structure and core functions while identifying key internal stakeholders. Like an analysis of important stakeholders and roles in the buying center at the customer, a comparable analysis can also be made internally to identify key relations. Building internal relations with colleagues in other functions such as marketing, human resources, purchasing, IT or R&D enhances Company Insight. Building an internal network enables the exchange of ideas and increases the understanding of the company competences and resources. It is important to build a broad network, and to build relations with people before we may need them, not at the moment we need them. Networks should not only be built with people based on what they know, but also on how open they are to learning and change. Like selecting the 'change agents' within the customer's organization it is also important to select and build relations with people 'open to change' within one's own organization. Building a broad internal network is the basis for connecting people to their counterparts in the customer organization.

Conducting Multi-Level Relations

To build new connections and encourage people to cooperate is the basis for moving from the so-called 'bow-tie' customer-supplier interaction, beyond the 'diamond' customer-supplier interaction towards an 'integrated' interaction (see for example McDonald et al. (2000), Dingena (2002) and Yip and Bink (2007b) on the different types of interactions). The relationship moves from a 'single point of contact' towards a situation in which people on multiple levels in both organizations and from different departments and functions across company boarders collaborate, thereby changing the role of strategic salespeo-

> **Orchestrating Customer-Supplier Interaction at Siemens**
>
> Siemens AG is a globally operating technology company with core activities in the fields of energy, healthcare, industry and infrastructure. The company employs 363,000 employees and had consolidated revenue of € 75.9 billion in 2013. Within Siemens, relationships with strategic customers are built on multiple levels. As the president of Siemens Information & Communications Corporate Accounts commented: 'Customers do not want a single point of contact, but rather a collaborative multifunctional team aligned to their organization'. For this collaborative, cross-functional team, customer focus means having a deep understanding of the customer and its industry sector, focusing on the customer's business processes, having knowledge of and selling all of Siemens' assets. Next to the cross-functional team, sector boards provide a total portfolio to meet the customer's needs (...) With this structure it is the whole corporation, not just the global account manager that interacts with the global customer'.
>
> Source: Siemens Annual Report (2013) and Yip and Brink (2007b, p. 13)

Exhibit 5.4 Orchestrating customer-supplier interaction at Siemens (Siemens Annual Report 2013; Yip and Bink 2007b, p. 13)

ple from being a 'linking-pin' between both organizations towards becoming a conductor of multi-level relations. As phrased by a strategic account manager: 'The account manager used to be like the gatekeeper. The success or failure of the account strategy relied on their credibility. Now they are more of an orchestrator; it's their job to get resource managers inside of the opportunity so they can internalize it themselves' (as cited in Helsing et al. 2003, p. 21).

By identifying and connecting key stakeholders at multiple levels in both organizations both planned and unforeseen customer challenges can be dealt with in such a way that joint value is created (Senn et al. 2013). In addition, the customer-supplier relationship becomes less dependent on the interactions between individual people. Building and nurturing these multi-level relations will create a basis for acting in the best possible way in the so-called 'Moments of Truth' (Carlzon 1987) in the customer-supplier interaction.

Elevating Sales to the Boardroom Agenda

Building the internal network includes elevating sales to the boardroom agenda and engaging C-level in the right way to boost performance. Senn (2006) observes both examples of effective use of senior management involvement and examples of under-leveraged or even counterproductive involvement of senior executives in strategic sales. One of the major findings of a study over the past decade (studying over 100 companies) is that senior executives' personal customer interactions will significantly increase

> **Leverage of Established Relations Within Hospitality Group in Moment of Truth**
>
> 'As a result of the 9/11 attack on New York's Twin Towers, one of the supplier's major customers lost, in one tragic day, its central office in the United States. The hospitality group itself was affected by continued travel bans and concerns. However, because of its relationship network and credibility with the customer and within its own organization, the global account director of the supplier firm was able to protect both companies' business by transforming hotel space and marshalling additional resources, including IT and reception services, into temporary office space for several hundred customer staff, within just a few days'.
>
> Source: Senn et al. (2013, p. 37)

Exhibit 5.5 Leverage of established relations within hospitality group in moment of truth (Senn et al. 2013, p. 37)

a company's growth rate (Senn 2006). 'In successful cases, sales and profits can double or even triple within two or three years because of a systematic, replicable executive engagement process' (Senn 2006, p. 28). Senn shares the result of the implementation of an executive sponsorship program at Siemens. Within Siemens this executive engagement process in strategic sales is called 'Top Executive Relationship Process' (TERP). In this process, strategic account managers are required to plan at least eight prepared executive engagements with their strategic customers per year. Figure 5.2 shows the impact of this process both for Siemens and for their strategic customers.

After a four year program an analysis was made of the impact of the executive engagement. The growth rate of the 'TERP' managed customers was compared to the growth rate of the 'non-TERP' managed customers. This resulted in a growth-rate factor two (see Fig. 5.3) for the accounts where top executives were actively engaged.

Strategic Internal Relationships May Provide Access to Cross Boundary Relations

Beyond building relations with internal stakeholders, strategic sales teams need to build cross boundary relations with relevant stakeholders in the created value network. These may include the company's suppliers or alliances within the value network. Piercy and Lane (2009) argue that the internal strategic relationships may play a vital role in getting access to cross boundary relations with company suppliers and alliances (see Fig. 5.4). 'It is likely that strategic internal relationships which will be vital to achieving effective

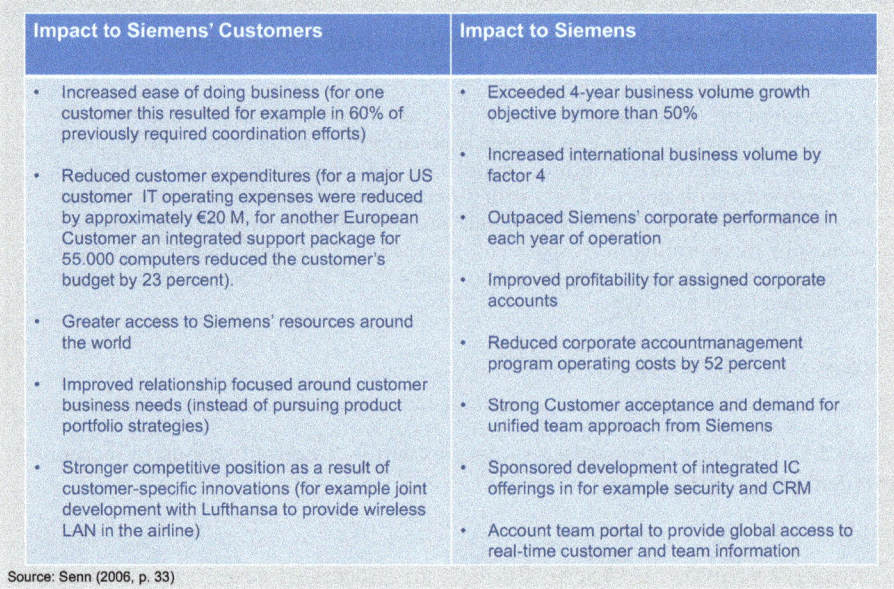

Fig. 5.2 Benefits of top executive engagement to Siemens and their strategic customers. Source: Senn, 2006, p. 33, reproduced with permission

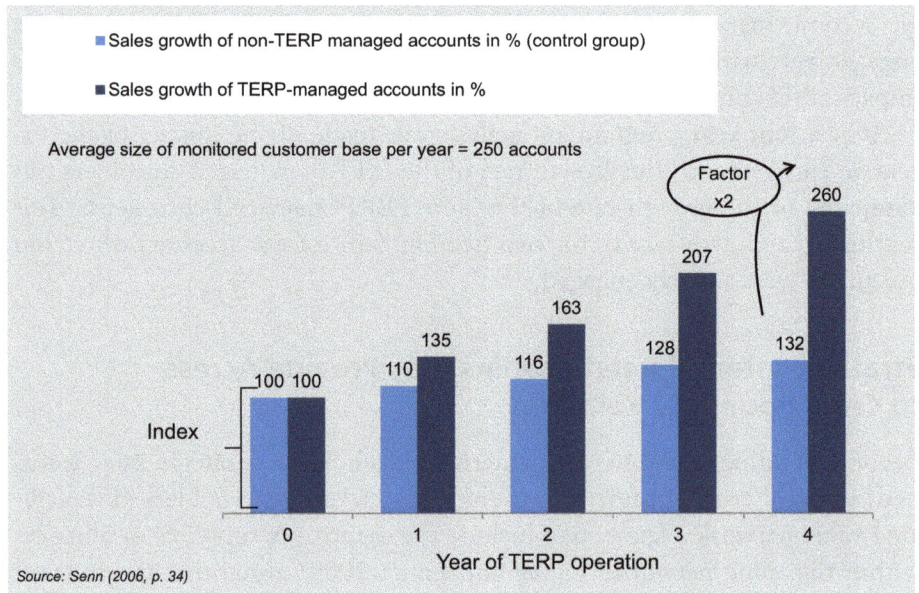

Fig. 5.3 Impact of Top Executive Relationship Process (TERP) at Siemens: The Executive Growth Factor, Source: Senn, 2006, p. 34, reproduced with permission

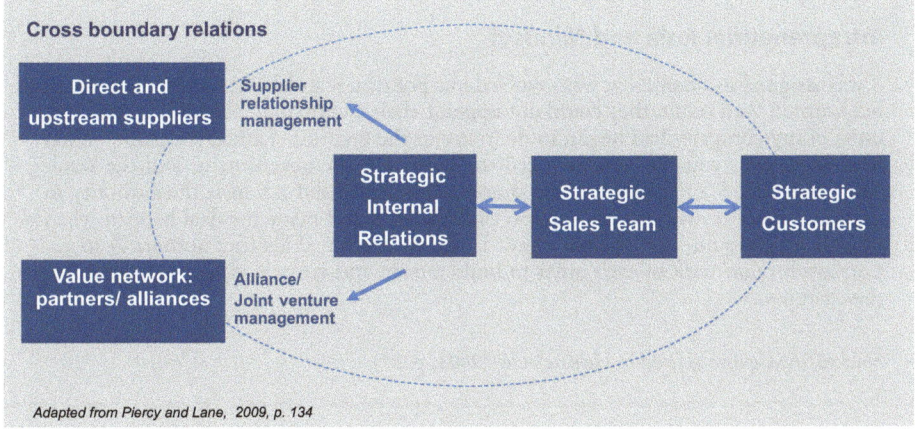

Fig. 5.4 Strategic internal relations may provide access to cross boundary relations

integration in networked companies will be between organizational units and processes that manage key external relationships (...) The management of that coordination will require the effective management of relationships between those responsible for strategic customer management, those who manage relationships with suppliers, and those who are tasked with the management of alliance and joint venture relationships with external organizations' (Piercy and Lane 2009, p. 134). In practice, people within the customer organization may have better relations and more knowledge of the seller's suppliers and alliances than individuals within the strategic sales team. As commented by Piercy (2010), the need is to align strategic customer management with strategic supplier management within the seller organization.

5.2 Creating Alignment Around Vivid and Factual Business Opportunities

It is vital to success to regard transformational sales as a business rather than as a sales initiative (see also Sherman et al. 2003). Transformational sales requires an *intra*preneurial role and mindset within strategic sales teams, who essentially become entrepreneurs within the boundaries of the organization. As defined by Eesley and Longenecker (2006, p. 19), 'intrapreneurship is propelled by an individual's or a team's willingness to take calculated risks and act to create business opportunities that serve an organization's need for growth and improvement'. In practice, this may require an attitude of rather 'asking for forgiveness than for permission'. In many cases transformational sales in-

> **Intrapreneurial Role and Mindset**
>
> 'I was assigned to a bank that was involved in a political scandal and was on the verge of bankruptcy. As a result, they could not upgrade their technology. The local operational units of my company had begun to de-invest in the account. I asked the bank, 'What if we took over your IT'. I convinced the bank and the government, and the result was a five-year, $ 230 million dollar outsourcing deal. I did not have the authority to do this. I went in without the support of leadership and made the deal happen when most people would have walked away. I used leadership skills (not authority) to get people who don't talk to each other to build a team and create a value proposition for the customer.'
>
> Source: SAM quote as cited in Helsing et al. (2003, p. 21)

Exhibit 5.6 Intrapreneurial role and mindset (Helsing et al. 2003, p. 21)

volves changing the rules of the game. Part of the success of strategic sales teams is the willingness to bend or sometimes break the rules. 'Most innovation involves breaking or bending rules. Not rule breaking that is unethical or dishonest, but rule breaking that is necessary to getting ideas designed, built and out the door' (McKeown 2008, p. 33).

In order to align internal stakeholders around the set transformation agenda and to mobilize the required competencies and resources, the actual business opportunity needs to be crystal clear. It is of eminent importance to vocalize, visualize and calculate the business opportunity.

Vocalize, Visualize and Calculate the Business Opportunity

Alignment around business opportunities requires painting a compelling picture of the envisioned outcome. To see and articulate not only what is there, but to additionally vocalize and visualize what can be. To make the business opportunity vivid to internal stakeholders, strategic sales teams need to vocalize, visualize and calculate the actual opportunity. The concept of personalized value propositions (Sect. 4.3) may also be applied to communicate and demonstrate the value internally, vocalizing the business opportunity in such a way that it 'resonates' internally, in a way that people can 'feel and understand' (Quinn 2004). To foster imagination, visual tools (pictures, movies, and storyboards) may also be used. At the same time the business opportunity has to be calculated in monetary terms building both an inspiring and fact-based business case in which the monetary value is calculated and demonstrated. Quinn (2004) uses the term 'grounded vision' to refer to the integration of both factual and imaginative elements. Crafting a grounded vision upon the

> **Electrolux Profit and Loss Statement for Strategic Retail Accounts**
>
> In order to measure the financial value created in the interaction with strategic retail customers, Electrolux Europe started with two measures of value creation: operating profit and asset utilization. The value creation Profit and Loss statement includes the following measures:
>
> - Volume and sales revenue,
> - Variable manufacturing costs,
> - Variable transactional costs,
> - Supporting costs to serve for the account,
> - Customer net assets tied up by the account (principally accounts receivable and inventory), multiplied by a weighted average cost of capital to reflect the carrying costs incurred by Electrolux Europe for carrying these assets in the books.
>
> Source: Bailey and Hesslschwerdt (2006)

Exhibit 5.7 Electrolux Profit and Loss Statement for strategic retail accounts (Bailey and Hesselschwerdt 2006)

actual opportunity may inspire others to contribute; or as Quinn states: 'The integration of reality and possibility creates an image that attracts self and others outside the comfort zone into a state of active creation' (Quinn 2004, p. 140).

Joint Profit & Loss Statement

Calculating the business opportunity can start with creating a Profit and Loss statement of the chosen value-innovation opportunity. Within Electrolux Major Appliances (the European Division of AB Electrolux an international household appliance company) Profit and Loss statements are created to assess the value of the interaction with their strategic retail customers (see Exhibit 5.7).

In a collaborative relationship with customers, it is preferable to design a 'Joint Profit and Loss Statement' to clarify the value for both the customer and the supplier. In addition to the joint transformation agenda (Appendix C), a joint Profit and Loss Statement can to be elaborated around the chosen value innovation opportunities. This includes the increased (economic and emotional) benefits in relation to the invested (economic and emotional) sacrifices both from the customer and from the supplier perspective. As the perceived value of doing business with the supplier should be larger as compared to the considered alternatives, the perceived value of doing business with

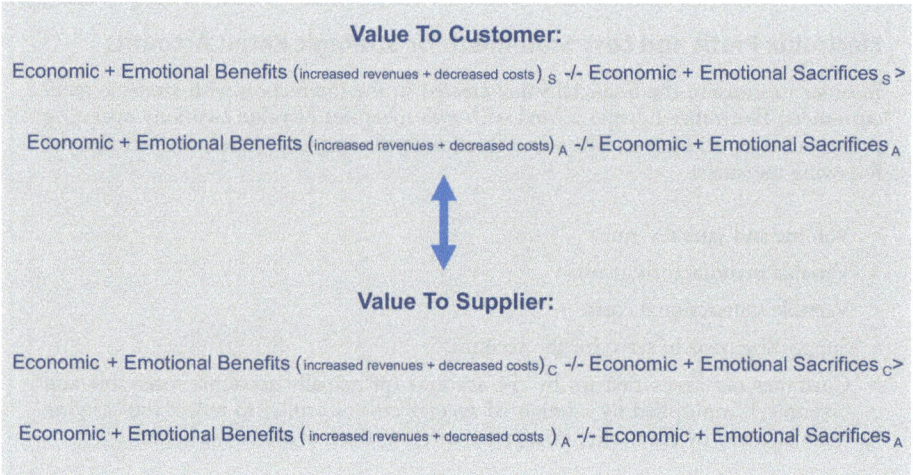

Fig. 5.5 Joint Profit & Loss

the customer should also be larger (or at least equal) as compared to considered alternative customers. Anderson et al. (2007, p. 62–63) suggest that a useful device for increased collaboration is a 'gives and gets analysis'. Even though not all emotional costs and revenues can be quantified in monetary terms, we propose to include these elements in the joint evaluation of value gained by both parties. To get an indication of increased trust levels between both parties, the 'Trust equation' of Maister et al. (2000) as discussed in Sect. 4.1 can be used.

As shown in Fig. 5.5, the *value to the customer* may be discussed by analyzing the economic and emotional benefits gained (increased revenues and/or decreased costs) in interaction with a particular supplier (S) minus the experienced economic and emotional sacrifices given in relation to this supplier and as compared to alternatives available (A). Likewise the *value to the supplier* may be analyzed by analyzing the value of the interaction with this particular customer (C) as compared to alternatives available (A). Appendix E provides a template that could be used as a starting point for defining joint longer term objectives and for making the actual joint 'Profit and Loss' statement more explicit.

Scalability: Indirect Value Derived From Joined Learning

It is important to include both the actual and potential value in the joint Profit and Loss Statement. This includes the indirect value that may be derived from the joined learning with our strategic customers. As mentioned by Martijn Legemaat and Swinda Hagedoorn (see example of TNT) in sales cases, it is

Creating Alignment and Delivering the Promise at TNT

TNT Express operates in 200 countries around the world. As Swinda Hagedoorn (Director Global Solutions Management, TNT) states: 'Within TNT an entrepreneurial sales approach is chosen in relation to our Top 150 Strategic Accounts worldwide. On a monthly basis we organize a session where Sales Managers can present their Sales Cases for approval. In this forum, which is attended by two Board members, the cases are challenged and if successful handed over to Innovation Management to start the design. The solution design follows an innovation process that is developed following the LEAN (**D**efine, **M**easure, **E**xplore, **D**evelop, **I**mplement DMEDI) methodology. Having a clear process in place helps to ensure stage gates are passed, all responsible functions are involved at the right time and the final business case is fact-based. Ultimately the process leads to a faster time to market of the solution, a solution that is co-developed with lead customers and that is fully accepted by all entities in the company, which greatly helps in the roll-out of the new solution.

Throughout the innovation process the customer plays an important part. We can delight our customers only if we fully understand the impact of our service on the customer's processes. We invite a key customer for a joined workshop to define the service. In this session we present our key processes, including operations, customer service, reporting and invoicing. We invite the customer to respond and tell us what happens in their own processes when we fail to deliver on these '*Moments of Truth*'.

We get interesting new insights from these sessions. For example, one customer wanted to prevent us from having any electronic communication with their customers (the end consumers). Our customer was afraid to lose touch with their end consumers, and therefore wanted to own all the messaging regarding the delivery of their consumer electronic products. TNT's messages to end consumers are triggered by automatic scans of parcels as they move through our network (like the scan of a parcel being loaded in a truck). These messages are close to real-time, whilst our customer has to translate TNT's messages into their own messaging systems and then send it to their end consumers. Valuable time is lost and mistakes often occur. The performance benefits of 'TNT' messaging were demonstrated by benchmarking this customer with other TNT consumer electronics customers, where TNT did the messaging. Only by engaging the customer in-depth in our process and by discussing the impact of our process on their business could we identify a true customer delight opportunity. In this case we identified a unique solution where TNT's standard messages were personalized, using the company name of the customer. As a result, the customer was happy with the personal touch and the service improved significantly. Personalized messaging is now a standard feature of TNT's B2C solution'.

Source: Interview with Swinda Hagedoorn (Director Global Solutions Management, TNT), June 2013

Exhibit 5.8 Creating alignment and delivering the promise at TNT (based on interview with Swinda Hagedoorn, Director Global Solutions Management TNT, June 2013; published with permission)

important to include 'scalability', explicitly assessing to which extent a particular value innovation opportunity can be relevant to other customers or other applications. In this way the learnings and revenues of investing with strategic customers may be further leveraged both within the supplier and within the customer organization.

5.3 Impact: Leading From Any Chair

In order to mobilize available and accessible capabilities, strategic sales teams need to have impact within their own company. 'Finding ways to have impact inside your own company is what will allow you to have more impact with your customers' (Helsing et al. 2003, p. 20). In their survey among strategic account managers, Helsing et al. ask what it looks like once people have impact within the own organization. A few of the quotes are summarized in Exhibit 5.9.

So how do strategic sales team members create impact? In addition to building internal networks and creating alignment around compelling business ventures, there is a third, more fundamental aspect in guiding change and enabling internal transformation. This is the conscious decision of strategic sales team members on their (explicitly or implicitly) chosen influencing strategy in effecting change. In his initial model, Quinn (1988) distinguishes eight influencing or leadership styles that 'change agents' may explicitly or implicitly use. Quinn (2004) distinguishes four general 'strategies for effecting change'. Building on the model of Quinn we would say the four strategies are based on two dimensions describing the explicit or implicit normative models or beliefs people may have regarding change:

- *Dimension 1: Agility or responsiveness to market change.* The possible beliefs underneath business agility or responsiveness to market change are based on the explicit or implicit assumption that we primarily need to *A) focus on preserving or maintaining what is currently there* (protecting or maintaining the current internal status quo) versus the belief that we *B) need to con-*

What Does It Look like when You Have Impact Inside Your Own Company?

- *'Everyone wants to work on my team; we have team spirit.'*
- *'I've created some kind of change and someone else understands my point of view.'*
- *'I mobilize cross-functional work teams to lend resources toward strategic relationships with high value accounts.'*
- *'Other functional organizations in my company support projects and activities for my team's customers.'*

SAM quotes as cited in Helsing et al. (2003, p. 20)

Exhibit 5.9 What does it look like when you have impact inside your own company? (Helsing et al. 2003, p. 20)

tinuously adapt to changing environments, in fact that we need to embrace (unforeseen) change (see also Taleb 2012).
- *Dimension 2: Assumption about evoking change in others.* The possible beliefs underneath evoking change in others may be the explicit or implicit assumption that *A) change is a result of exerting control* versus the belief that *B) change is a result of releasing untapped potential or possibility.*

The four strategies for effecting change as distinguished by Quinn (2004, p. 73) are: telling, forcing, participating, and transcending. It is worthwhile for strategic sales teams to reflect on the (explicit or implicit) change strategy chosen before trying to drive internal change.

The *telling (which we refer to as 'convincing') strategy* is based on facts and rational persuasion. Change agents using this approach will try to explain and convince colleagues why their contribution is relevant and necessary. They may ask themselves questions like: 'Am I within my expertise?', 'Have I gathered all facts?', 'Have I done a good analysis?', 'Will my conclusions withstand criticism?', 'Are my arguments clear?', 'Are people listening?' (Quinn 2004, p. 73). The major assumption of this change strategy is that others will change once they understand why this change is needed and once they are convinced of the chosen route.

The *forcing strategy* (which we refer to as *'imposing'* strategy) is based on exerting authority. Change agents using this approach will make use of the formal power that is assigned to them. In fact they 'dictate' the kind of change they wish to see. They may ask themselves questions like: 'Is my authority firmly established', 'Is the legitimacy of my directive clear', 'Am I capable of imposing sanctions?', 'Is there a clear performance-reward linkage', 'Am I using maximum leverage?', 'Are the people complying?' (Quinn 2004, p. 73). The major assumption of this change strategy is that others will change once they are told to do so by people with formal authority.

The *participating (which we refer to as 'bridging') strategy* is based on relationships and open dialogue. Change agents using this approach will try to incorporate the ideas of stakeholders involved. They may ask themselves questions like: 'Is everyone included in an open dialogue?', 'Do I model supportive communication?', 'Is everyone's position being clarified?', 'Am I surfacing the conflicts?', 'Are the decisions being made in a participative way?', 'Are the people cohesive'? (Quinn 2004, p. 73). The major assumption of this change strategy is that others will change once they are involved in the change process.

The *transcending (which we refer to as 'inspiring') strategy* is based on the possibility people have to transcend or excel, inspiring others to release untapped potential. Change agents using this approach will try to focus upon the willingness of people to contribute and grow. They may ask themselves questions

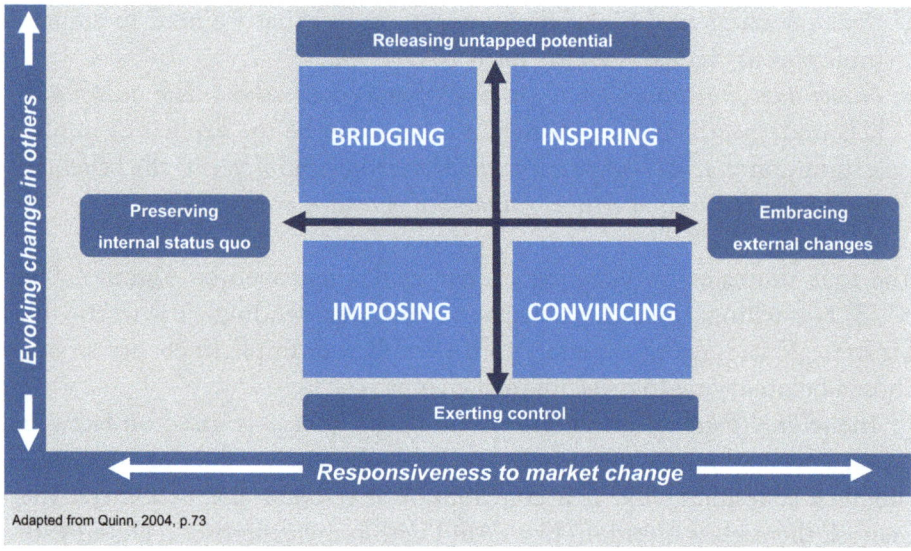

Fig. 5.6 Strategies for driving internal change

like: 'Is my purpose clear?', 'Am I other-focused', 'Am I able and willing to see the potential in others?', 'Am I externally open, moving forward into uncertainty?', 'Are the people walking with me into uncertainty?' (Quinn 2004, p. 73). The major assumption beyond this change strategy is that others will change once they get (better) in touch with their own untapped potential.

In Fig. 5.6 an adapted version of Quinn's model (Quinn 2004, p. 73) is depicted which can be used by strategic sales team members to consciously decide on their chosen strategy in effecting change.

Inspiring to Release Untapped Potential Internally as Well

In a transformational sales context we believe that the 'convincing strategy' (see Fig. 5.6) won't help much to really align people and mobilize the necessary resources. Colleagues may well be willing to listen to all the arguments involved but will not necessarily feel the urge to change behavior themselves. In practice, the 'imposing strategy' is not often applicable since in most cases there is not a formal authority relation between strategic salespeople and the colleagues they try to align. A bridging strategy which includes colleagues in conversations can be useful. This strategy enables to connect, to listen and to include what is important to others. The most profound impact can be realized by inspiring others, empowering others to realize their full potential and actively inspiring colleagues to contribute their best to new business opportu-

> **The Silent Conductor**
>
> 'I had been conducting for nearly twenty years when it suddenly dawned on me that the conductor of an orchestra does not make a sound. His picture may appear on the cover of the CD in various dramatic poses, but his true power derives from his ability to make other people powerful. I began to ask myself questions like: 'What makes a group lively and engaged?' (…) In the light of my 'discovery' I began to shift my attention to how effective I was at enabling the musicians to play each phrase as beautifully as they were capable'.
>
> Source: Zander and Zander (2000, p. 69)

Exhibit 5.10 The silent conductor (Zander and Zander 2000, p. 69)

> **How Much Greatness Are We Willing to Grant?**
>
> 'A monumental question for leaders in any organization to consider is: How much greatness are we willing to grant people? Because it makes all the difference at every level who it is we decide we are leading. The activity of leadership is not limited to conductors, presidents and CEO's of course – the player who energizes the orchestra by communicating his newfound appreciation for the tasks of the conductor, or a parent who fashions in her own mind that her children desire to contribute, is exercising leadership of the most profound kind'.
>
> Source: Zander and Zander (2000, p. 73–74)

Exhibit 5.11 How much greatness are we willing to grant? (Zander and Zander 2000, p. 73–74)

nities. Strategic sales team members can consciously consider a question raised by Zander and Zander in their publication 'Leading from any chair' (Zander and Zander 2000): 'how much greatness are we willing to grant people?' Benjamin Zander, a music conductor of the Boston Philharmonic, describes his discovery of what he refers to as 'the silent conductor' (Zander and Zander 2000; see Exhibit 5.10).

For strategic sales team members the shift to becoming 'the silent conductor' may require a shift in focus from what 'we need from others' to 'what they may be capable and willing to contribute'.

Inspirational leaders (as inspirational strategic salespeople) are able to connect and touch upon people's passion.

The most inspiring strategic salespeople are those who do not only guide transformation with external customers, but who additionally enable transformation within their own organization, guiding change by touching upon and listening for passion and commitment and empowering colleagues in different

> **Listening for Passion**
>
> Listening for passion and commitment is the practice of the silent conductor whether players are sitting in the orchestra, on the management team, or on the nursery floor. How can this leader know how well he is fulfilling his intention? He can look in the eyes of the players and prepare to ask himself, 'Who am I being that they are not shining?' He can speak to their passion. He can invite information and expression. He can look for an opportunity to hand them the baton.
>
> (Zander and Zander 2000, p. 74)
>
> Today was exceptional in that I learned leadership is not a responsibility (…) Things change when you care enough to grab whatever you love, and give it everything.
>
> (Amanda Burr, student at The Walnut Hill School as cited in Zander and Zander (2000, p. 74))

Exhibit 5.12   Listening for passion (Zander and Zander 2000, p. 74)

parts of the organization to realize their full potential. In fact they are leading change.

> The **most powerful way to increase impact** may be to inspire others to release untapped potential. This requires a shift in focus from '*what we need from others*' towards '*what they may be capable and willing to contribute*'.

References

Anderson, J. C., Kumar, N., & Narus, J. A. (2007). *Value Merchants*. Boston: Harvard Business School Press.

Bailey, P., & Hesselschwerdt, P. (2006). How aligned is your organization on the needs and methods for creating value? *Velocity Strategic Account Management Association, 8*(2, Q2), 15–19.

Carlzon, H. (1987). *Moments of Truth*. Cambridge: HarperCollins Publishers.

Dingena, M. (2002). *Key Account Management*. Deventer: Kluwer.

Dingena, M. and C. Teven, 2015, 'Top 7 Challenges in Strategic Sales Practice', Internal Paper, Marketing Planning Centre Nederland/Rotterdam school of Management.

Eesley, D. T., & Longenecker, C. O. (2006). Gateways to Intrapreneurship. *Industrial Management, 48*(1), 18–23.

Grönroos, C. (2007). *Service Management and Marketing. Customer Management in Service Competition*. Chichester: John Wiley & Sons.

Helsing, J., Ceraghty, B., & Napolitano, L. (2003). *Impact without Authority*. Chicago: Strategic Account Management Association.

Maister, D. H., Green, C. H., & Galford, R. M. (2000). *The Trusted Advisor*. New York: The Free Press.

McDonald, M., Rogers, B., & Woodburn, D. (2000). *Key Customers. How to manage them profitably*. Oxford: Butterworth-Heinemann.

McKeown, M. (2008). *The Truth about Innovation*. Upper Saddle River: Pearson, Prentice Hall.

Piercy, N. F. (2010). Evolution of Strategic Sales Organizations in Business-to-Business Marketing. *The Journal of Business &amp Industrial Marketing*, *25*(5), 349–359.

Piercy, N. F., & Lane, N. (2009). *Strategic Customer Management*. Oxford: Oxford University Press.

Quinn, R. E. (1988). *Beyond Rational Management: Mastering the Paradoxes and Competing Demands of High Performance*. San Francisco: Jossey-Bass.

Quinn, R. E. (2004). *Building The Bridge as you walk on it*. San Francisco: John Wiley & Sons.

Senn, C. (2006). The Executive Growth Factor: How Siemens Invigorated its Customer Relationships. *The Journal of Business Strategy*, *27*(1), 27–34.

Senn, C., Thoma, A., & Yip, G. S. (2013). Customer-Centric Leadership: How to Manage Strategic Customers as Assets in B2B Markets. *California Management Review*, *55*(3), 27–59.

Sherman, S., Sperry, J., & Reese, S. (2003). *The Seven Keys to Managing Strategic Accounts*. New York: McGraw-Hill.

Siemens Annual report 2013.

Speakman, J. I. F., & Ryals, L. (2012). Key Account Management: The Inside Selling Job. *Journal of Business &amp Industrial Marketing*, *27*(5), 360–369.

Taleb, N. N. (2012). *Antifragile*. London: Penguin Books.

Yip, G. S., & Bink, A. J. M. (2007b). *Managing Global Customers*. Oxford: Oxford University Press.

Zander, R. S., & Zander, B. (2000). *The Art of Possibility*. New York: Penguin Books.

6

Undertaking the Transformative Journey

> *Excellence never lies within the boxes drawn in the past.*
> *To be excellent, leaders have to step outside the safety net of*
> *the company's regulations [...] And to bring about deep*
> *change in others, people have to reinvent themselves*
>
> Quinn (1996, p. 11)

Transformation of the customer-supplier interaction is a process of leading change. It does not happen automatically. It can be regarded as a 'transformative journey' (Johnson and Fillipini 2009, as cited in Wießmeier et al. 2012). In practice the customer and supplier will go through different stages of collaboration (Cordón and Vollmann 2008) before they enter a stage of strategic alignment. 'What is needed is a joint transformation; a combined commitment to striking out in a new direction and abandoning the current ways of working' (Cordon and Vollman 2008, p. 24). The increased levels of coordination cannot be mandated but instead need 'to be nurtured over time by a collaborative mind-set and behavior' (Wießmeier et al. 2012, p. 21). This requires a learning partnership at all touch points in the customer-supplier interaction (see Sect. 6.1). The joint transformative journey to value innovation involves taking calculated risks and dealing with uncertainty. To a certain extent it is a journey into the unknown which requires to 'pave the path' or 'build the bridge' as we walk on it (Quinn 1996, 2004). We will discuss the principles of Quinn (2004) that may be useful traveling this road (see Sect. 6.2). We will end this book with where, in fact, it all begins. We believe that guiding business change starts with a conscious reflection upon the own (implicit) assumptions that strategic salespeople may have. We will summarize some of the assumptions that we observe in practice and that may be challenged to drive change and make a difference (see Sect. 6.3).

6.1 Creating a Learning Partnership at All Touch Points

Transformative customer-supplier relationships are built over time. Essential for a growing collaborative relationship is that both organizations support each other in their growing and learning processes. This requires a learning partnership at all 'touch points' between both organizations. These touch points may include both personal interactions and virtual interactions, for example using (mobile) Internet technologies to enable virtual (cross-organizational) teams to work effectively together. Different players in the value network may be included in team meetings. In order to create a learning partnership, the actual touch points in the customer-supplier interaction must be known. In a study including 18 large organizations operating both in the business-to-consumer and business-to-business sectors, Payne et al. (2008, p. 85) distinguish three elements in the 'process-based value co-creation framework between customers and suppliers'. These are:

- *Customer value-creating processes*, which in a business-to-business context can be referred to as the processes which the customer organization uses to manage their business and their relationships with suppliers.
- *Supplier value-creating processes,* which are the processes, resources and practices which the supplier uses to manage its business and its relationships with customers and other relevant stakeholders.
- *Encounter processes* and practices of interaction and exchange that take place within the customer and supplier relationships.

In Fig. 6.1 the relationship experience between the customer and supplier is depicted, based on the conceptual framework of Payne et al. (2008, p. 86).

Within the encounter processes, significant interaction moments or customer-supplier customer-supplier touch points can be identified. As argued by Carlzon (1987), these touch points may be seen as 'Moments of Truth' in which actual value may be delivered or destroyed. In collaborative relationships the actual Moments of Truth can be analyzed in a joint effort between customer and supplier. This may be the starting point not only for creating a 'flawless execution' (Cordón and Vollmann 2008) between both parties, but also the starting point for increased learning and trust within the relationship, uncovering which moments in practice really matter and where suppliers are really able to make a difference. Many important issues can only be discovered through learning by doing. Customer-supplier relationships really become 'transformative' once both companies are committed to sharing honest feedback and wish to learn.

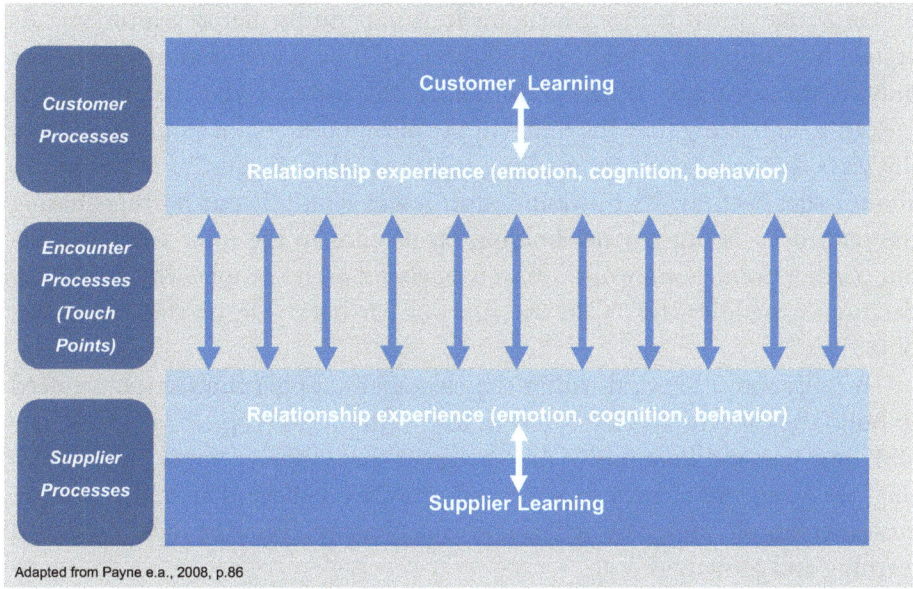

Fig. 6.1 Relationship experience: learning at all touch points

According to Cordón and Vollmann (2008) collaborative customer-supplier relations go through four stages: starting at 'flawless execution' as a prerequisite for further collaborative development, moving on towards a situation in which total cost of ownership can be realized, towards total customer value requiring a long-term collaboration (for example including joint R&D and innovation) towards a situation of 'strategic alignment' including alignment at senior management level.

Shared Improvement Agenda and 'Joint Performance Dashboard': Learning Real-Time

Based on their shared learning, transformative customer-supplier relations have to improve continuously to adapt to and drive change within their markets. To evaluate supplier performance, procurement departments commonly assess supplier performance, for example using a 'supplier quality index', allowing purchasing managers to compare the quality performance of different suppliers (Van Weele 2014, p. 298). Within transformative relations however, instead of a 'one-way' assessment of supplier performance, preferably a two-way 'joint performance dashboard' is created. This dashboard can include observations shared by various people within both the customer and supplier organization. Experiences in the interaction in the defined 'Moments of Truth' may provide input for joint learning and further development. As argued by Cordón and Vollman (2008, p. 110), it is important to focus not only on measuring of factual performance but also to add 'perception based measures' indicating how the different people in both organizations feel about the ways the companies are working together. They suggest to ask opinions about issues such as: 'Is this joint effort a win-win?' 'Is our partner sharing the rewards?' 'Is our partner holding up its ends in the joint work?' 'Does our partner do its homework?' 'Can we depend on its people?' 'Does it make decisions expeditiously?' 'Can we trust our partner?' 'Do we share common objectives?'

To really move forward within the customer-supplier interaction we need to build the capacity to 'learn in real time' (Quinn 2004, p. 153). In analogy with Spreitzer and Porath (2012), who argue that employees will 'thrive' once they are given the chance to learn and grow, we believe that transformative customer-supplier relations will thrive once there is a genuine focus on joined learning and growth as well.

> **Within transformative relations** instead of a 'one-way' assessment of supplier performance, **preferably a two-way 'joint performance dashboard' is created.** This dashboard can include observations shared by various people within both the customer and supplier organization

6.2 Paving the Path as You Walk on It

The joint transformative journey to value innovation involves taking calculated risks and dealing with uncertainty. To a certain extent it is a journey into the unknown which requires to 'pave the path' or 'build the bridge' as we walk on it (Quinn 1996, 2004). This journey can be supported by strategic customer management programs embedded within the supplier organization. In a study on global customer management programs, Capon and Senn (2010) mention that developing the capabilities to build effective global customer management is a long-term process that requires a gradual implementation approach, working gradually from a 'pilot stage', via a 'springboard stage' towards an 'embedded stage'. They argue that in practice there is no short cut or easy answer. We also believe that developing transformative customer-supplier relationships will be most successful following this gradual implementation approach, starting with a limited number of strategic customers, building learning experiences and adapting approaches accordingly. Even though supporting processes and systems will inevitably enhance the success of transformational sales, it is at the same time important to keep it as simple as possible, acknowledging the intrapreneurial role and mindset required to successfully drive change with strategic customers. As mentioned by Ivo Rook (Director Northern Europe at Vodafone Global Enterprise): 'Real value is created beyond systems and processes' (see Exhibit 6.1).

Fostering transformative customer-supplier relations goes beyond processes and systems. It is in essence about how effective both parties are at leading change. As Quinn (2004) remarks, deep (or disruptive) change requires developing leadership. It is about understanding leadership and 'developing it in ourselves or in others'. Quinn (2004) describes 'eight creative states' that enable the development of leadership. We believe these 'creative states' are a prerequisite to building transformative customer-supplier interactions. Within this chapter we would like to highlight and explore in depth five of these creative states that we believe are fundamental to building transformative customer-supplier interactions. These are: reflective action, authentic engagement, appreciative inquiry, grounded vision, and tough love.

Reflective Action in Customer-Supplier Interactions

> Most of us in the organizational world (…) are engulfed in action, at the expense of contemplation and reflection. This extreme is just as isolating as the extreme of contemplation divorced from action (Quinn 2004, p. 99).

As remarked by Quinn, in the corporate world we often become 'addicted to action'. Before understanding the real customer business challenges, we have already come to an answer or conclusion of how to solve them. In particular, sales teams skilled in 'solution selling techniques' may have a tendency to be engulfed in action and possible solutions. Whilst it is a major strength to be fast in finding solutions to existing problems, transformative insights result rather from reflecting upon the factual challenge encountered.

> In particular sales teams skilled in 'solution selling techniques' may have a tendency to be engulfed in action and possible solutions. Whilst it is a major strength to be fast in finding solutions to existing problems, **transformative insights result rather from reflecting upon the factual challenge encountered.**

As argued by Quinn (2004), reflective action is about the integration of being active and energetic while at the same time being reflective. Within a

Vodafone: The Power of Simplicity

Vodafone Group Plc is one of the world's leading telecommunications groups, with a significant presence in Europe, the Middle East, Africa and Asia Pacific. The company's group revenue amounts € 57.3 billion (year ended 31 March 2014). Vodafone supports its global customers to create 'a competitive edge' using its global technology and networks and has seen considerable growth with these enterprise customers. The basis of Vodafone's success in the enterprise customer segment is the dedicated business unit 'Vodafone Global Enterprise' (VGE). As Ivo Rook (Director Northern Europe at VGE) states: 'Since the creation of VGE its revenues have grown in 6 consecutive years outperforming the market and every other of Vodafone's sales channels. Within these 6 years VGE has generated € 2.7 billion revenues for Vodafone. Additionally VGE is recognized by their customers and analysts such as Gartner and IDC as the leader in Global Managed Mobility. Starting point to success is a deep understanding of the customer business and understanding how your capabilities and services can contribute to the customer's business objectives.

Supporting customers to better run their business and serve their customers in a better way
In essence VGE transforms global customers' businesses. We help our global customers like Amazon, BAE Systems Applied Intelligence, Bosch and Luxottica to better manage their costs and improve operational agility. We also support our customers to keep ahead of the game, helping them to get closer to their customers and improve the customer experiences in their markets. Together with global customer TomTom, we are improving driver experience by enabling TomTom to provide real-time traffic data to its end users across different countries. The collaborative partnership with Amazon resulted in an improved user experience of Amazon customers' worldwide [see the earlier example of the Kindle e-reader in Sect. 4.2]. With Bosch we partner in electric mobility. Electric mobility is a new way of mobility and requires thinking about charging stations and parking. The charging stations need to offer new services such as authentication of the driver, reservation, monitoring of the charging process and all the information needed for the billing. The charging unit contains a corresponding communications unit and the Vodafone SIM card is used to activate the communication.

Solid bidding process and business case
Building global solutions requires a careful, fact based gating process, reflecting all elements of the global solution (jointly) created with global customers:

- Are all costs (fixed and incremental) known and embedded in the business case?
- Are all operational processes present to service the solution?
- Are resources and capabilities available to deliver the services?
- Are all operational, financial and legal risks known and mitigated?
- Are there major dependencies on 3rd parties or externally obtained technology?

Building such a process and adhering to it rigidly is crucial to deliver a consistent service experience and to enforce the right behavior across the globe. Take time to build the bid-process and ensure its outcome is acknowledged across the group. If it is well executed it will underpin the global mandate which is essential to grow the business.

> *Use intelligence for creating (fact based) value, keep the execution as simple as possible*
> To be successful, I challenge everyone who guides business transformations with global customers to be ambitious, but to keep it simple. The mission for VGE was to build a globally managed mobility portfolio, to create a structure that enables building transformative services with global customers in a single contract and to build a sales force and service organization which supports the actual transformation. My personal experience and advice: keep the complexity limited to the strategic choices involved in the selection and creation of your global portfolio and global value propositions, but keep the execution of your global sales program as simple as possible:
>
> - Chose one sales-and-bid process and adhere to it rigidly. There are many tools available which manage a sales process. They all aim to build consistency around how sales teams approach their strategic customers and how teams can capture their efforts to make a strategic sale. In essence these methodologies don't differ much, their major goal is to drive a consistent approach and speak the same language. So chose one and don't allow other methodologies to exist alongside it. Make sure the bid-process is fact based (see above).
> - Chose one supportive CRM-tool that's easy to use. Some providers have a tendency to try and capture every aspect of your business in their system and drive deep integration even into your production processes. My advice is to resist any deep integration which goes beyond your sales process, it will increase the complexity of a global roll-out and it will make your system rigid and expensive. A good CRM-tool should be easy to use, even easier to access and is always available. If executed well the system will become the main communication source and then it delivers true value: global coordination of a fast growing and very successful global workforce.
> - Serve your strategic customers from a dedicated unit. Build a separate, dedicated unit for strategic enterprise customers which has a commercial and operational mandate and sits across your operating companies.
>
> Real value is created beyond systems and processes. So leave these tools and processes take their course, don't try to control it too much. Nothing can go wrong, since you have built the right portfolio, you sell transformative value, the bids are approved through a consistent process and you capture all activities in a single CRM-tool to which you have on-line access, that's what we call the power of simplicity.'
>
> Source: Interviews with Ivo Rook (Director Northern Europe at Vodafone Global Enterprise) in the period April–September 2014, and Vodafone (2013, 2014)

Exhibit 6.1 Vodafone the Power of Simplicity (based on interviews with Ivo Rook, Director Northern Europe at Vodafone Global Enterprise, April to September 2014, and Vodafone (2013, 2014); published with permission)

customer-supplier relationship this means being deeply engaged in the customer's business situation and at the same time spending time reflecting upon possible growth and learning, thereby keeping track of observations within the customer-supplier encounter process, joint reflecting on the earlier mentioned 'Moments of Truth' and joint reflection upon possible (emerging) future worlds. Instead of selling to our customers we have to strategize with them. The real value may not be to come up with solutions, but instead to jointly think strategically and come up with the right questions and insights to broaden perspectives. As commented by McKeown (2012, p. 11), the essential value of strategic thinking as compared to 'just working hard' is that it 'requires you to question what is being done and what could be done'. This is about 'opening your mind to possibilities. It's about seeing the bigger picture. It's about understanding the various parts of your business, taking them apart, and then putting them back together again in a more powerful way. It's about insight, invention, emotion and imagination focused on reshaping some part of the world' (McKeown 2012, p. 15). Reflective action in customer-supplier interactions is the starting point for joint learning and growth.

> **Reflective Action in customer-supplier interactions**: instead of selling to our customers, we have to think strategically with them, using insight and disrupting current ways of thinking to reshape the customer and supplier business as it is known today.

Authentic Engagement in Customer-Supplier Interactions

> Those with huge positive influence understand the power of relationships, connection, and engaging (…) they're not afraid to get 'out there' connecting with others, sharing their knowledge and talents, offering their authentic and often contrarian viewpoints (…) they understand that positive, supportive and authentic relationships are foundational building blocks to anything and everything they want to achieve (Caprino 2014).

Deepening the customer-supplier relationship requires the willingness of both parties to move beyond transactions. To build a transformative relationship and make a difference requires the combination of on the one side being involved, connected and committed to a shared purpose and on the other side remaining authentic. This involves the willingness of both parties to bring in their own perspective and maintain integrity, close the gaps, explore and combine both self-interests and the interests of the other. In essence they need to have a genuine interest in guiding and enabling business transformation and need to be passionate about making a difference. As argued by Quinn (2004, p. 114) this means 'bringing in an evolving and authentic self to a passionately held shared purpose'. This starts with the professional passion of those involved. What is it that people really care about? In practice real differences in customer-supplier interactions are made once people share a professional passion and are truly able to shift perspectives. As argued by Byrd and Chamberlain: 'Self-actualizing people are, without one single exception, involved in a cause outside their own skin, in something outside themselves' (Byrd and Chamberlain, as cited in Quinn 2004, p. 110). Authentic engagement in customer-supplier interactions requires that partners 'love what they do' (Quinn 2004, p. 114) and share a passion to jointly make a difference.

> **Authentic engagement in customer-supplier interactions** requires a shared passion of both parties to jointly make a difference and genuinely contribute their best to a shared purpose – even though they work in different firms.

Appreciative Inquiry in Customer-Supplier Interactions

Once customers and suppliers are authentically engaged in a shared purpose, this also opens up room for joint learning and business change in a more fundamental way. As argued by Quinn (2004, p. 135) it is important to 'recognize that there is untapped energy in every person, relationship, and system'. And in order to release this untapped potential we have to remove ourselves 'from the expert role and take the role of positive inquirer'. Business transformation requires an understanding of assumptions driving behavior within customer and supplier organizations. Transformational sales is about challenging the customer and exploring possibilities to create and provide new perspectives upon these (very often implicit) assumptions. As argued by Dixon and Adamson (2011, p. 52–53) this is the 'power of insight' (...) speaking directly to 'an urgent need of the customer not to buy something but to learn something'. We would like to add that this power of insight is applicable to the sales organization as well, speaking to the need of the supplier not only to sell something but to learn something as well, enabling learning both within the customer organization and within the supplier organization.

This learning may be fostered by choosing an approach of 'appreciative inquiry' in customer-supplier relations. The concept of appreciative inquiry was originally introduced by Cooperrider as 'the capacity to see the most creative and improbable opportunities in the marketplace (...) the capacity to see with an appreciative eye the true and the good, the better and the possible...' (Cooperrider, as cited in Quinn 2004, p. 127). As commented by Quinn (2004, p. 125–126), it is an 'integrative state' combining being 'constructively optimistic at the same time that we are realistic and analytic' (...) it is the kind of constructive questioning that surfaces what people care most about, inviting their 'commitment and releasing energy and creativity'. In addition to joint planning, appreciative inquiry supports reacting in the best possible way to unplanned events. As argued by McKeown (2012, p. 27) 'The smart strategist allows strategy to be shaped by events (...) Evidence supports the idea that the most successful entrepreneurs and leaders are fantastic at noticing opportunities. And the greatest opportunities come from reactions to unplanned events'. A similar observation was made by Johnson et al. (2012, p. 28) 'Significantly we found that alternative leaders were able to accelerate the pace of transformation, not by forcing the issue but by leveraging what we call happy accidents to gain a broad platform of support. Happy accidents are unanticipated circumstances or events that ultimately support transformation in the direction favored by the leaders-in-waiting'.

> **Appreciative Inquiry in customer-supplier interactions** enables both parties to recognize the untapped potential, both within the customer-supplier relationship and within the customer and supplier business. Being able to focus on the best of what currently is, seeing the possibility in unforeseen market events and nourishing the imagination to see what future business could look like.

Grounded Vision in Customer-Supplier Interactions

Guiding business transformation within the customer and supplier organization requires an articulated and 'grounded' vision of the possible future. As elaborated in the Chaps. 4 and 5 both to guide customer business transformation and to enable the transformation within the supplier organization, the vision needs to be vocalized in a way stakeholders can 'feel and understand' (Quinn 2004, p. 138). A resonating value proposition (both towards external and internal stakeholders) combines both the perspective upon the possible impact or outcome (compelling vision) and the untapped business potential (prioritized business challenge – 'grounded in reality'). As argued by Quinn (2004, p. 140), 'radical change requires to move to the root and to ground a vision in 'lived experience'. A grounded vision is 'grounded and factual' while also hopeful and visionary' [conveying] a 'future that emerges from the realities of the existing system (...) The integration of reality and possibility creates an image that attracts self and others outside the comfort zone and into a state of creative action'. A grounded vision touches upon the unspoken needs that are beneath existing business challenges and headaches and attracts both people in the customer and in the supplier organization out of their comfort zone.

> A **grounded vision of the customer-supplier relation and the future business** may inspire people to move beyond current routines, get rid of 'existing boxes' and actually release the previously unrecognized business potential. By jointly choosing this innovative approach the customer-supplier collaboration moves beyond existing competition and defines a different playing field.

Tough Love in Customer-Supplier Interactions

Transformative customer-supplier relations are not created in a short time. Van Weele (2014, p. 207) argues that successful partnerships in buyer-seller relationships are in the minority. 'The few successful partnerships in buyer-seller relations do not come as a surprise. Cooperation with suppliers requires internal teamwork between all the disciplines involved. The functional structure in many companies interferes with an effective internal cooperation and as a consequence interferes with a close and effective cooperation with suppliers'. Cordón and Vollmann (2008, p. 63) argue that the best performance within the customer-supplier relationship is realized when parties treat each other with 'tough love', illustrating this concept with a trade-off between too little and too much pressure within the relationship. 'If you apply excessive pressure, you will become unattractive, and the supplier will reduce its efforts and try to exit the relationship. If there is very little pressure, the customer-supplier team may become complacent. Of course as is true of employees, some collaborative efforts need no external pressure to deliver high performance – they apply it themselves'.

Applying 'tough love' in customer-supplier relations is combining 'being assertive and bold yet compassionate and concerned' (Quinn 2004, p. 186). In guiding business transformations, 'tough love' will enable people both in the customer and supplier organization to learn, stretch and grow. To challenge each other in a genuine way, focused on increasing excellence and greatness, increasing not only the productivity of the relationship, but also the competitiveness and success in the market. 'Those who treat me with tough love disturb the habitual way in which I choose to see myself by asking me tough questions and making tough statements. (…) to provide the integration of tough love that empowers people to move forward. During deep change, people have to move outside the comfort zone and learn new behaviors. This means surrendering control, and no one wants to do it. At such times people need both purpose and support. That is what tough love does' (Quinn 2004, p. 187). In customer-supplier interactions it is about challenging and supporting each other in such a way that both parties are able to make a real difference within the marketplace.

> **Tough Love in customer-supplier interactions** results in a challenging relationship in which both parties encourage each other to live up and work to their full potential.

6.3 Ending Where It All Begins: Challenging the Own Assumptions

As argued by Caprino (2014), people with impact commit to 'continually bettering themselves', they 'have an openness to see, learn and experience new things'. Transformational sales requires disrupting both the customer's and the supplier's thinking and assumptions about their business. To be able to make this transformation happen also requires a conscious reflection upon one's own (implicit) assumptions. In fact we are ending this book where it all began. In this last section we would like to summarize ten sales assumptions that we observe in practice and which have been highlighted throughout this book. These assumptions may hamper building transformative customer-

10 Sales assumptions hampering transformative relationships

1. 'Large customers are strategic customers'
2. 'Past customer performance predicts future potential'
3. 'We are of strategic importance to customers who are of strategic importance to us'
4. 'Articulated customer needs are the starting point for value creation'
5. 'We need to focus on stakeholders with the 'authority to spend'
6. 'In the end, all that counts for the customer is lowering the price'
7. 'Superior value propositions focus on the main points of difference as compared to competitive alternatives'
8. 'Sales is (or should be) an 'outside job' and the real fun is in building external relations (with our customers)'
9. 'Mobilizing internal resources and competencies requires a focus on what we need from others'.
10. 'I am able to make real progress once the right systems and processes are in place'

Fig. 6.2 Challenging the own assumptions

supplier relations. This list is not meant to be exhaustive; it is rather meant as starting point for discussion, an aid to reflect upon and challenge the assumptions strategic salespeople may have themselves (see Fig. 6.2).

As Quinn (2004) argues, all business transformation starts with personal transformation. Rephrasing a powerful quote of Gandhi he says: 'When we change ourselves, we change how people see us and how they respond to us. When we change ourselves, we change the world' (Quinn 2004, p. 24). We wish you an inspirational journey, driving change with strategic customers and guiding life-changing and company-altering transformations.

References

Citing Sources

Capon, N. and C. Senn, 2010, 'Global Customer Management Programs: How to Make Them Really Work', *California Management Review*, Vol. 52, Issue 2, pp. 32–55.

Caprino, K., 2014, '9 Core behaviors of people who positively Impact the World', *Forbes.com*, 2 June, 2014.

Carlzon, H., 1987, *Moments of Truth*. New York: HarperCollins Publishers.

Cordón, C. and T. E. Vollmann, 2008, *The Power of Two. How Smart Companies Create Win-Win Customer-Supplier Partnerships That Outperform the Competition*. Basingstoke: Palgrave MacMillan.

Dixon, M. and B. Adamson, 2011, *The Challenger Sale. Taking Control of the Customer Conversation*. New York: Portfolio/Penguin.

Johnson, G., G.S. Yip and M. Hensmans, 2012, 'Achieving Successful Strategic Transformation', *MIT Sloan Management Review*, Spring 2012, Vol 53, No. 3, pp. 24–32.

McKeown, M. 2012, *The Strategy Book*. Harlow: Pearson Education Limited.

Payne, A.F., Stobacka, K. and P. Frow, 2008, 'Managing the Co-Creation of Value', *Journal of Academic Marketing Science*, Vol. 36, Issue 1, pp. 83–96.

Quinn, R.E., 1996, *Deep Change: Discovering the Leader within*. San Francisco: John Wiley & Sons.

Quinn, R.E., 2004, *Building The Bridge as you walk on it*. San Francisco: John Wiley & Sons.

Spreitzer, G. and C. Porath, 2012, 'Creating Sustainable Performance', *Harvard Business Review 90*(1), pp. 93–99.

Vodafone, 2013, Annual report 2013.

Vodafone, 2014, Vodafone Global Enterprise. Amazon Case Study, published on Vodafone website.

Weele, A.J. van, 2014, *Purchasing and Supply Chain Management. Analysis, Strategy, Planning and Practice*. Andover: Cengage Learning.

Wießmeier, G.F.L., Thoma, A. and C. Senn, 2012, 'Leveraging Synergies between R&D and Key Account Management to Drive Value Creation', *Research-Technology Management*, Vol. 55, Issue 3, pp. 15–22.

Further Reading

Caniëls, M. and C. Gelderman, 2005, 'Purchasing strategies in the Kraljic matrix: a power and dependence perspective', *Journal of Purchasing and Supply Management*, 11 (2-3), 141–155.

Lusch, R.F., Vargo, S.L. and M. O'Brien, 2007, 'Competing Through Service: Insights from Service-Dominant Logic', *Journal of Retailing*, Vol. 83, Issue 1, pp. 5–18.

Montgomery, D.B. and G.S. Yip, 2000, 'The Challenge of Global Customer Management', *Marketing management*, Vol. 9, Issue 4, pp. 22–29.

Pine, B.J. and J.H. Gilmore, 1998, 'Welcome to the experience economy', *Harvard Business Review* 76(4), pp. 97–105.

Ritter, T., 2006, 'Communicating Firm Competencies: Marketing as Different Levels of Translation', *Industrial Marketing Management*, Vol. 35, Issue 8, pp. 1032–1036.

Rozemeijer, F.A., Weele, A. van, and M. Weggeman, 2003, '*Creating Corporate Advantage Through Purchasing: Toward a Contingency Model*', The Journal of Supply Chain Management, Vol. 39, Issue 1, pp. 4–13.

Storbacka, K., 2012, 'Strategic Account Management Programs: Alignment of Design Elements and Management Practices', *Journal of Business & Industrial Marketing*, Vol. 27, Issue 4, pp. 259–274.

Taleb, N. N., D.G. Goldstein and M.W. Spitznagel, 2009, 'The Six Mistakes Executives make in risk management', *Harvard Business Review* 87(10), pp. 78–81.

Tanner, R. 2014, *Organizational conflict: Get used to it and use it*. May, 13, 2014, Amazon Digital Services.

Terho, H., Haas, A., Eggert, A. and W. Ulaga, 2012, '*It's Almost Like Taking the Sales Out of Selling – Towards a Conceptualization of Value-Based Selling in Business Markets*', Industrial Marketing Management, Vol. 41, Issue 1, pp. 174–185.

Thomas, K.W. and R. Kilmann, 2002, *Thomas-Kilmann Conflict Mode Instrument*. Mountain View: CPP Inc.

Valk, W. Van der and F. Wynstra, 2012, 'Buyer-Supplier Interaction in Business-to-Business Services: A Typology Using Case Research', *Journal of Purchasing & Supply Management*, Vol. 18, Issue 3, pp. 137–147.

Walter, A., Ritter, T. and H.G. Gemünden, 2001, '*Value Creation in Buyer-Seller Relationships*', Industrial Marketing Management, Vol. 30, Issue 4, pp. 365–377.

Weele, A.J. van, Rozemeijer, F.A., 1996, 'Revolution in purchasing', *European Journal of Purchasing & Supply Management,* vol. 2, issue 4, p. 153–160

Wilson, K., T. Millman, D. Weilbaker and S. Croom, 2001, *Harnessing Global Potential. Insights into Managing Customers Worldwide.* Chicago: Strategic Account Management Association.

Zeithaml, V.A., 1988, 'Consumer Perceptions of Price, Quality and Value: A Means-end model and Synthesis of Evidence', *Journal of Marketing*, 52 (3), pp. 2–22.

Appendix

Appendix A: Separating Strategic Customers from Others

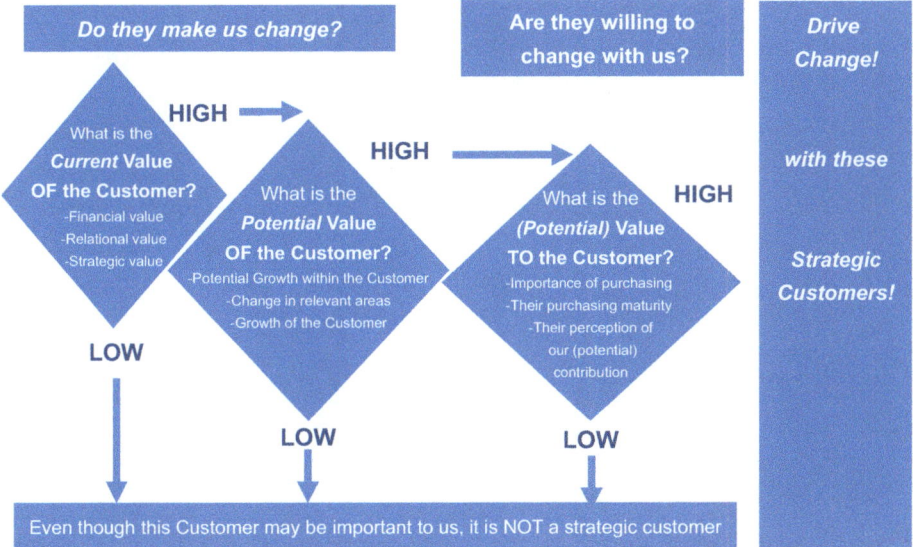

Appendix B: Differentiated Customer Strategies

Differentiated Customer Strategy & Service Level	Transactional Customers	Development Customers	Large Customers	Strategic Customers
❏ Strategy				
❏ Customer Interface & Interaction				
❏ Customer Intelligence & (Joint) Planning				
❏ Service level				
❏ % Time/ resources				

Appendix C: Joint Transformation Agenda

Joint Transformation Agenda Customer/ Supplier:	Joint Vision and Strategic Objective:		
Insight *Top 3* -Customer Business Challenges	1.	2.	3.
-Customer Business Headaches	1.	2.	3.
-Supplier Available Capabilities	1.	2.	3.
-Supplier Accessible Capabilities	1.	2.	3.
Strategic focus *Top 3 strategic focus points for value innovation (Customer/Supplier collaboration matrix may be used to prioritize focus points)* 1. <Competency/resource to be used/ developed or acquired to address < Customer Business Challenge or Business Headache> 2. <Competency/resource to be used/ developed or acquired to adrdress < Customer Business Challenge or Business Headache> 3. <Competency/resource to be used/ developed or acquired to address < Customer Business Challenge or Business Headache>			
Top 3 Personalized Value Propositions (selected stakeholders 'open to change')			
1. Person (prioritized stakeholder)	Untapped Potential:	Perspective:	Reason to belief:
2. Person (prioritized stakeholder)	Untapped Potential	Perspective:	Reason te belief:
3. Person (prioritized stakeholder)	Untapped Potential:	Perspective:	Reason to belief:
Implementation *Top 3 Actions in Customer Organization* 1. Who (within customer dmu)	Why?	What?	When?
2. Who (within customer dmu)	Why?	What?	When?
3. Who (within customer dmu)	Why?	What?	When?
Top 3 Actions in Supplier Organization 1. Who (colleague within supplier dmu)	Why?	What?	When?
2. Who (colleague within supplier dmu)	Why?	What?	When?
3. Who (colleague within supplier dmu)	Why?	What?	When?

Appendix D: Joint Profit & Loss Statement

Perspective		Joint Profit & Loss Statement			
		Increased revenues		Decreased costs	
		Economic revenues	Emotional revenues	Economic costs	Emotional costs
Customer	Direct				
	Indirect (scalability)				
Supplier	Direct				
	Indirect (scalability)				

Appendix E: How Transformational is your Sales? – Self-assessment

Building Block 1: Driving Change with Strategic Customers

Transactional				Transformational
Customer differentiation				
We do not differentiate between large and strategic customers				Strategic customers are separated from large customers
1.	2.	3.	4.	5.
Match or mismatch?				
Customers are prioritized based on their importance to the supplier organisation				Customers are prioritized based on the importance of the customer and the importance to the customer
1.	2.	3.	4.	5.
Value logic				
Turn product into cashflow				Make customer more successful in their markets
1.	2.	3.	4.	5.
Supplier (Sales) Mindset				
Customer is in a 'state of certainty'				Customer is in a 'state of uncertainty'
1.	2.	3.	4.	5.
Major customer reason for interaction with supplier				
To Buy				To Learn
1.	2.	3.	4.	5.

Transformational Sales

Building Block 2: Setting the Joint Transformation Agenda

Transactional → Transformational	1.	2.	3.	4.	5.
Starting Point for Value Creation	Articulated Customer Needs				Insight into upcoming Customer Business Challenges in their markets
Focus within Customer Buying Center	Primary focus on decision makers and stakeholders with the 'authority to spend'				Primary focus on stakeholders 'open to change' and being able to influence decision makers
Required Company Insight (supplier organization)	Requires understanding of existing products and services				Requires understanding of all available and accessible capabilities within the supplier organisation and value network
Account planning process	Task of accountmanager/ accountteam				Joint customer /supplier process resulting in joined transformation agenda
Planning Horizon	<1 year				>2-3 years

Building Block 3: Guiding Customer Business Transformation

Transactional →		Transformational
Supplier purpose		
1. Differentiate by demonstrating value as compared to competitive offerings	2. 3. 4.	5. Make a Difference by broadening the customer perspective on its business moving beyond competition
Customer Value Proposition		
1. USP of product or solution	2. 3. 4.	5. Vision on the release of untapped customer business potential
Supplier impact		
1. In the end all that counts for the customer is lowering the price	2. 3. 4.	5. In the end we are able to increase customer competitiveness and success in their markets by decreased economic or emotional costs and/or increased economic or emotional revenues
Belief upon how to 'score' with customers		
1. First and foremost we have to convince our customers with rational facts and figures.	2. 3. 4.	5. To be able to guide the customer in making their business run better we have to be able first to genuinely connect on an emotional level with prioritized stakeholders
Belief upon how to convey a value proposition		
1. We have to sell our proposition to the customer in a compelling way	2. 3. 4.	5. To bring a value proposition to life requires to enable the customer to see the potential that's there, yet unseen

Building Block 4: Enabling Internal Transformation

Transactional → Transformational				
Sales Mindset				
1. Sales is (or should be) an 'outside job' and the real fun is in building external relations with customers	2.	3.	4.	5. Business transformation requires an integrative perspective upon the sales role
Perception of strategic customer management within supplier organisation				
1. Sales initiative (focus on revenues)	2.	3.	4.	5. Business Venture (Joint P&L)
Sales team role				
1. Selling	2.	3.	4.	5. Being an intrapreneur within the boundaries of the organisation
Customer-Supplier interaction				
1. Single point of contact (bow-tie interaction)	2.	3.	4.	5. Cross boundary/ or integrated interaction on multilevels in both organisations, including C-level
Belief upon impact/ effecting change				
1. Mobilizing internal resources and competencies requires a focus upon what we need from others	2.	3.	4.	5. In order to increase impact we need to inspire others to release untapped potential. This requires a focus upon what others may be capable and willing to contribute

Appendix

Building Block 5: Undertaking the Transformative Journey

Transactional → Transformational				
Performance Measurement				
1. There is a 'one-way' assessment of supplier performance by the customer	2.	3.	4.	5. In order to monitor performance a 'joint (customer/supplier) performance dashboard' is created
Change process				
1. Once we have a clear roadmap about how to implement the proposed solutions at the customer place, we are halfway there	2.	3. Impact	4.	5. The transformative journey to value innovation with our strategic customers is to a certain extent a journey into the unknown, which requires to 'pave the path' as we walk on it
1. We solve customer problems	2.	3.	4.	5. In the end we transform both the customer and supplier business
Learning				
1. We regularly ask our customer for their feedback	2.	3.	4.	5. We foster and encourage learning at all touch points within the customer-supplier interaction
Business transformation				
1. I am able to make real progress once the right systems and processes are in place	2.	3.	4.	5. Real value is created beyond systems and processes. All business transformation starts with personal transformation

Credit Lines for Permission Clearance

Exhibit 2.2: Assessing the customer relationship, Based on Senn (2012, p. 38), reproduced with permission.

Exhibit 2.3: Joint Innovation with strategic Automotive Customers at Kendrion. Based on interviews with Dr. Bernd Gundelsweiler (CEO Division Automotive) and Piet Veenema (CEO Kendrion) in September 2013, published with permission.

Figure 2.4: International Purchasing Survey: purchasing ratios across industries. IPS Data, 2009, provided by Finn Wynstra, Rotterdam School of Management, reproduced with permission.

Figure 2.5: Dupont-analysis Heineken NV (2014): impact of purchasing savings on Return on Capital Employed, Van Weele, 2014, p. 13, updated by Van Weele with data 2014 in April 2015, reproduced with permission.

Figure 2.6: Six stages of purchasing maturity and related purchasing focus, Based on Van Weele 2014, p. 68, reproduced with permission.

Exhibit 2.9: Deepening the understanding of purchasing strategies: include competitive priorities. Source Ateş (2014), and interview with Melek Ateş Mach 2014, reproduced with permission.

Figure 3.2: BCG's value creators report: the global population is increasingly connected. Source: Boston Consulting Group 2013: The 2013 TMT Value Creators Report: The Great Software Transformation, reproduced with permission.

Exhibit 3.2: Industry 4.0: the fourth industrial revolution is already on its way, Roland Berger, 2014, p. 7–9, reproduced with permission.

Exhibit 3.4: Royal DSM: Customer Insight means 'thinking B-to-C and acting B-to-B'. Based on interviews with Mauricio Adade (Chief Marketing Officer DSM, Theo Verweerden (Marketing Program Director Value Creation, Rossana Rodriguez (Senior Marketing Consultant, DSM) in November 2014, Company Presentation 2014, DSM at a Glance, DSM Factbook, published with permission.

Exhibit 3.5: New Lens Scenarios at Shell, Extract from the 'New Lens Scenario's publication, Shell international, 2013', reproduced with permission.

Exhibit 3.6: Festo: Embedded engineers at Marel. Based on interviews with Folkert Hettinga (Industrial Sales Manager Food & Beverage, Agriculture at Festo), April 2014, and Festo Highlights 2014. Published with permission.

Exhibit 3.7: Europcar and Daimler: Car2go-on-demand mobility. Based on interview with Esther van Koot (Commercial Director Europcar Netherlands) in May 2014 and Europcar Activity Report 2011–2012, published with permission.

Exhibit 3.8: Philips: applying natural daylight simulation technology in promising areas. Based on interviews with Selin Kelleci-Van Balen (Senior Regional Product Marketing Manager at Philips Lighting), Matthew Cobham (Lighting Application Team Manager, Indoor Professional Lighting Solutions Europe), June 2014 and Philips Annual report 2013, Philips 2013 (Schoolvision), Philips 2014, Lighting Europe/AT Kearney, 2013, published with permission.

Exhibit 3.9: ABInBev and JF Hillebrand: redefined value in Global Beverage Logistics, Based on interviews with Pierre Bonel (Chief commercial Officer) and Sander Ouwehand (Corporate Accountmanager), December 2013–April 2014, published with permission.

Exhibit 4.3: Europcar moving your way: flawless experience for business travelers, Based on interview with Esther van Koot (Commercial Director Europcar Netherlands and Europcar Activity Report 2011–2012, published with permission.

Exhibit 4.5: Festo: Reducing Total Cost of Ownership for their global customers, Based on interview with Folkert Hettinga (Industrial Sales Manager Food & Beverage, Agriculture at Festo), April 2014, published with permission.

Exhibit 4.9: Joint-Go-to-Market; Vodafone and Amazon increase 'always on experience'. Source: Vodafone 2014 – Vodafone Global Enterprise Amazon Case study, published on Vodafone website, Reproduced with permission.

Exhibit 4.11: Value-bridge at TNT: design a close to damage free process, Based on interview with Hugo Koppelaars, Director Sales TNT, February 2013, published with permission.

Exhibit 5.2: From Customer Insight to solid business development at TNT, Based on interviews with Martijn Legemaat, Corporate Account Insight Director at TNT, June 2013–January 2014, published with permission.

Figure 5.2: Benefits of top executive engagement to Siemens and their strategic customers. Source: Senn, 2006, p. 33, reproduced with permission.

Figure 5.3: Impact of Top Executive Relationship Process (TERP) at Siemens: The Executive Growth Factor, Source: Senn, 2006, p. 34, reproduced with permission.

Exhibit 5.8: Creating alignment and delivering the promise at TNT, Based on interview with Swinda Hagedoorn, Director Global Solutions Management TNT, June 2013, published with permission.

Exhibit 6.1: Vodafone: The Power of Simplicity, Based on interviews with Ivo Rook, Director Northern Europe at Vodafone Global Enterprise, April to September 2014, and Vodafone (2013, 2014), published with permission.

About the Authors

Philip Kotler (M.A., University of Chicago, Ph.D., M.I.T.) is the S. C. Johnson Distinguished Professor of International Marketing at the Kellogg School of Management, Northwestern University. He has published *Marketing Management*, 15th edition, *Principles of Marketing*, 16th edition, *B2B Brand Management, Ingredient Branding, Building Global Biobrands, Winning Global Markets* and 50 other books. His research covers strategic marketing, innovation, industrial marketing, and corporate social responsibility.

He has consulted GE, IBM, Apple, Honeywell, Ford, Merck, Samsung and many other companies and has lectured on all the continents. He has lectured to many companies about how to apply sound economic and marketing science principles to increase their competitiveness and growth. He has also advised foreign governments on how to develop the service quality of government agencies and how governments can support their domestic companies to prosper in the global marketplace. He has also extensively consulted nonprofit organizations on marketing strategies and policies.

In 2013, Professor Kotler was selected as the first recipient of the William L. Wilkie American Marketing Association Foundation's (AMAF) 'Marketing for a Better World' Award for significant contributions to marketing's theory and practice. Professor Kotler is the recipient of 22 honorary degrees from abroad. He has been chosen by the marketing profession as a Legend in Marketing.

Dr. Marian Dingena is visiting faculty at the Rotterdam School of Management (Erasmus University Rotterdam) and other European Business Schools and founder of MPCN Action Learning. At the Rotterdam School of Management she is involved in custom and open enrollment programs with corporate clients, such as the Strategic Account Management and the Sales Leadership Diploma Program. As a change management expert, Marian has over 20 years of international experience in guiding business transformation through action learning programs and customized interventions. She has experience across a wide range of industries and worked throughout Europe, Southern Africa, North America, and India.

Marian is specialized in Strategic Customer Management, Sales and Market Leadership and Change management. Marian is working as a business coach, sparring partner, lecturer, source of inspiration, and independent researcher.

Earlier publications include: *The Creation of Meaning in Advertising* (1994), *Successful Marketing Planning* (co-author, original publication: 1997), and *Key Account Management* (2002).

Customers repeatedly report having experienced a business-altering, career- and even life-changing impact as a result of her action learning programs and business coaching interventions.

Marian: *'It's my professional passion to explore genuine customer value and to inspire business leaders to make a significant and meaningful difference in the contribution of their choice'.*

About the Authors

Dr. Waldemar A. Pfoertsch is professor for International Business at the Pforzheim University, Germany. From 2007–2010 he was professor of marketing at China Europe International Business School Shanghai (CEIBS). His other teaching positions have been at the Executive MBA Program at the University of Illinois, Chicago and at the Steinbeis University in Berlin, the University of Cooperative Education Villengen-Schwenningen. He was visiting Associate Professor at Kellogg Graduate School of Management, Northwestern University and Lecturer for Strategic Management at Lake Forest Graduate School of Management. He has taught online with the University of Maryland University College Graduate School. At the start for his career he was Research Assistant at the Technical University of Berlin.

Waldemar Pfoertsch has extensive experience in management consulting. In his years at UBM/Mercer Consulting Group, Arthur Andersen Operational Consulting and LEK Consulting, he worked throughout Europe, Asia and North America, assisting companies in developing international strategies. His earlier positions include sale and strategy positions at SIEMENS AG in Germany/USA and being an Economic Advisor to the United Nations Industrial Development Organization (UNIDO) in Sierra Leone, West Africa.

He is the author of various books and numerous articles. The most current book was published with Katherine Xin, Arthur Yeung, and Shengjun Liu – *The Globalization of Chinese Companies: Strategies for Conquering International Markets*. *Ingredient Branding: Making the Invisible Visible* and *B2B Brand Management* were co-authored with Philip Kotler from Kellogg Graduate School of Management. He also published *Business-to-Business Marketing* with Rob Vitale and Joe Giglierano in 2010 and with Peter Godefroid *B2B Marketing* in 2009. In the field of Internet Marketing, he has published *Living Web and Internet Strategy*, books on application of Internet marketing and Internet strategy. He has also written numerous articles on international strategies; B2B Brand Management, Ingredient Branding, Internet Marketing, CRM and market opportunities in emerging markets. He holds various board positions with private and not-for-profit organizations.

Bring Us in to Speak at your Next Event

A keynote address to your staff assembly can be one of the most effective ways of spreading the newest insights of transformational sales. Whether you need one of our team to motivate your sales force, provide an executive briefing with your senior staff, or give your entire organization a new sense of what's possible in your highly competitive world, we can offer illuminating stories and practical advice on refining your marketing approach and the processes involved.

Are your market challenges more complex? Would a half-day or day-long workshop help you and your team work through the various angles of the transformation?

We also provide training workshops for small and large executive trainings at your facilities or at university settings in the Americas, Europe and Asia.

Let's discuss how a ***Transformational Sales workshop*** can help you build on your current strategic and tactical marketing plans.

You can contact us at www.transformationalsales.info

We look forward to getting in touch.

Index Key Words

5 forces   47, 48
21st century purchasing organization   23

A

Adaptive Capabilities   x, 3, 55, 56, 59
Adaptive capabilities   36
Anti-fragility   79
Asset utilization   22, 58, 75, 82, 111
Authority to spend   36

B

Balance of power   27, 29
Balanced relationship   25, 27
Big Data   44, 45, 47
Boardroom agenda   2, 32, 106
Bottleneck category   28
Buffer stock   29
Business altering value proposition   x, 3, 89–92, 94, 96
Business transformation   x, xiv, 2, 3, 5, 6, 35, 37, 66, 67, 71, 72, 74, 77, 90, 93, 104, 127–131, 133
B-to-B   50, 52, 147
B-to-C   50, 52, 147

C

CarPlay   49
Challenger Sale   2
Chief Marketing Officer   52, 147
Collaborative buyer-seller relationships   24
Commercial orientation   20, 22
Company insight   3, 41, 55, 56, 59, 62, 105
Competition   xi, 2, 34, 41, 48, 62, 89, 131
Corporate social responsibility   24
Co-creating value   48
Cross boundary relations   107, 109
Customer business challenges   2, 36, 41–43, 50, 55, 56, 59, 62–65, 78, 87, 93–95, 101, 125
Customer business headaches   64
Customer business transformation   5, 71, 72, 74, 77, 90, 130
Customer experience   xiv, 76, 78, 127
Customer insight   x, xi, 2, 3, 41, 42, 52, 54, 55, 92, 103, 147, 148
Customer organization   3, 17–19, 32, 33, 55, 57, 59, 72, 73, 75, 77–79, 89, 91, 92, 94, 97, 105, 109, 113, 122, 129
Customer portfolio   ix, 9
Customer relationship   xv, 9, 13, 14, 147
Customer/Supplier collaboration matrix   63
Cyber-physical system   47

D

Decentralization   47
Decision Making Unit (DMU)   x, xv, 3, 21, 33, 50, 52, 94, 104
Development customer   15, 33
Downstream customers   48, 85
Dupont analysis   19, 147
Dynamic experience economy   1

E

Economic costs   74, 75

Economic revenues   76
EDI system   21
Electromagnetic systems   15
Embedded employees   57
Emotional connection   72, 77, 79, 80, 91
Emotional costs   72, 74, 75, 78, 112
Emotional revenues   72, 74, 77
Energy efficiency   47
E-catalogue   29
E-ordering   29
E-payment   29
Evoking change in others   115
Executive engagement   xv, 107, 108, 149
Extended Decision Making Unit   x, 50, 52
External integration   20, 21

F

Food ingredient   48, 50, 52, 78
Forefront customer   ix
Four windows of opportunity   73, 74, 77
Fuel efficiency   15
Functional silos   xiii
Future potential   9, 11, 13, 34

G

Game changing asset   xiii
Guide business transformation   xiv, 5, 34, 67

H

Holistic approach   xiii, xiv, 1, 5

I

iCar   x, 47, 49
Industry 4.0   45, 47, 147
Industry trends   43, 47, 103
Inside selling role   102
Internal change   115, 116
Internal integration   20, 21
Internal transformation   3–5, 96, 101, 114

Intrapreneurial perspective   36
Intrapreneurial role   x, 3, 102, 109, 110, 125
Investment   18
Investments   1, 9, 11, 14, 58
iOS   49
iPhone   49

J

Joint innovation   14, 15, 23, 27, 31, 49, 65, 96, 147
Joint innovation process   15
Joint performance dashboard   123, 124
Joint profit & loss statement   111
Joint strategic focus   x, 41, 62–64
Joint transformation agenda   2, 3, 41, 63, 67, 87, 111

L

Large customer   9, 11, 16, 33, 75
Lead collaborators   x, 3, 72, 87
Learning partnership   xi, 4, 5, 121, 122
Learning process   ix, 122
Leverage category   28
Levi's 501   52
Life sciences   52

M

Making a difference   ix, 34, 35, 128
Marketing 3.0   xiv, 1
Meso   42, 43, 47
Monetary value   12, 110
Money making logic   72, 92

O

Outside-in understanding   36

P

Partnership   xi, 4, 5, 10, 16, 23, 24, 27, 29, 52, 66, 74, 86, 121, 122, 127, 131
Perspective   ix, 1–4, 6, 10, 12, 17, 21, 25–30, 34, 35, 37, 42, 43, 50,

63–65, 71, 72, 74, 79, 80, 82, 92, 94–96, 102–104, 111, 128–130
Phygital world   45
Profit & Loss   111, 112
Purchasing coordination   20
Purchasing maturity   20, 22, 24, 147
Purchasing portfolio   17, 25–27, 29, 32
Purchasing professionalism   20, 28
Purchasing ratio   17, 18, 32, 147
Purchasing risk   25, 26, 28
Purchasing task   20
Purchasing value   17, 25, 26, 28, 32

Q

Quarterly numbers   xiii

R

Radical transformation   2
Reciprocal value   x, 76, 87, 94
Reinvent   2, 29, 33, 89, 121
Relationship value   12, 13
Revenue generating capacity   73, 76, 84
Reverse marketing   24
Reverse purchasing   25, 27, 29
Risk   ix, 2, 4, 9, 25, 27, 28, 34, 53, 63, 64, 71, 75, 76, 78, 79, 109, 121, 124

S

Scalability   103, 112, 113
Service centric company   xiii
Share of wallet   11
Silent conductor   4, 102, 117, 118
Smart robots   47
Solution design   ix, 113
Sourcing transformation   18, 19
Specialized purchasing companies   29
Strategic customer   ix, x, 1, 2, 4, 5, 9–11, 14–17, 22, 32, 34, 41, 48–50, 52, 57, 59, 62, 65, 75, 81, 87, 94, 101, 102, 104, 106–109, 112, 124, 127, 133, 149
Strategic value   12–14

Supplier adaptive capabilities   x, 55, 56
Supplier adaptive capabilities grid   x
Supply risk management   24
Switching costs   25, 28, 78

T

Top executive relationship process   xv, 107, 108, 149
Total cost of ownership   x, xv, 20–22, 24, 28, 32, 75, 81, 83, 123, 148
Total customer value   20–22, 24, 123
Transactional customer   15, 33, 34
Transformational sales   ix, xi, xiii, xiv, 2–5, 34–37, 54, 72, 89, 94, 96, 101, 104, 109, 116, 125, 129, 132, 155
Transformative collaboration   2
Transformative journey   xi, 4, 5, 121, 124
Transformative relationship   17, 22, 30, 32, 34, 128
Trust   12, 13, 23, 62, 71, 73, 77, 79–81, 112, 122, 124
Trust equation   81, 112

U

Ultra-High Molecular Weight Polyethylene   52
Untapped business potential   42, 54, 72, 89, 90, 94, 130

V

Value based sourcing   24
Value bridge   72, 93, 148
Value chain integration   20–22
Value chain management   24
Value innovation   x, 1–3, 5, 14, 41, 57, 62, 63, 74, 87, 88, 96, 111, 113, 121, 124
Value innovation opportunities   x, 2, 3, 5, 41, 63, 111
Value network   x, 3, 34, 41, 55, 59, 62, 66, 72, 87–89, 107, 122
Value perspective   30

Value proposition   x, 3, 72, 82, 84, 89–96, 103, 110, 127, 130

Value to the customer   xiii, 10, 16, 17, 29, 30, 32, 41, 59, 71, 90, 96, 112

Virtual industrialization   47

W

Win-win   x, 14, 15, 94–96, 124

Index List of Companies

A

ABB   xiv, 1, 9, 22, 65, 96
ABInBev   x, 59, 62, 148
Amazon   ix, x, 85, 86, 127, 148
Ansell   52
Apple   x, 47, 49

B

Bayer   52
BMW Group   49
Boeing   x, 57, 84
Bombardier   xiv, 9, 23

D

Daimler   x, 58
Dell   x, 57, 75
Dyneema   52

E

Electrolux   xi, 111
Europcar   x, 58, 76, 148

F

Ferrari   49
Festo   x, 57, 82, 83, 148
Ford Motor Company   xiii

G

GE   x, xiv, 1, 9, 48, 66, 84
General Motors   49

H

Honda   79
HP   xiv, 1, 9
Hyundai Motor Company   49

I

IBM   xiv, 9, 18, 19, 48, 73, 74

J

Jaguar   49

K

Kendrion   15, 147
Kia Motors   49
Kodak   x, 86, 87

L

Land Rover   49
Levi Strauss & Co   52
LSI Logic Corporation   x, 85

M

Marel   x, 57, 148
McKinsey   23
Mercedes-Benz   49
Mitsubishi Motors   49

N

Nissan Motor Company   49

P

P&G – Proctor & Gamble   xiv, 1, 9, 89
Philips   x, 60, 65, 148
PSA Peugeot Citroën   49

R

Reebok   52
Royal DSM   50, 52, 88, 147

S

SAP   xiv, 1, 9

Shell   x, 45, 50, 53, 148
Siemens   xi, xv, 65, 87, 106–108, 149
Sony   24
Subaru   49
Suzuki   49

T

TNT   x, xi, xv, 93, 101, 103, 112, 113, 148, 149

Toyota Motor Corp   49
Trumpf Group   48

V

Vodafone   ix, xiv, 1, 9, 85, 86, 125, 127, 148, 149
Volvo   49

Lightning Source UK Ltd.
Milton Keynes UK
UKHW02n1330090418
320737UK00002B/36/P